J. S. Randall

Minerals of Colorado

J. S. Randall

Minerals of Colorado

ISBN/EAN: 9783744791281

Printed in Europe, USA, Canada, Australia, Japan

Cover: Foto ©berggeist007 / pixelio.de

More available books at **www.hansebooks.com**

MINERALS

of

COLORADO.

————— ▬ —————

J. S. RANDALL.

·

————— ▬ — — —

GEORGETOWN, COLO.
GEORGETOWN COURIER PRINT.
1887.

INTRODUCTORY.

In the preparation of this glossary, the compiler has depended largely upon the works of J. D. DANA for a description of the mineral species, except in cases where the local mineral has undergone examination. He has also obtained much valuable information from papers upon particular minerals, read before scientific associations, and from articles prepared for the scientific press.

From a mineralogical point of view the wealth of Colorado exceeds that of the other States. In the metallic minerals almost every variety has been found here. And doubtless many interesting minerals have been overlooked from causes that it is impossible to guard against. The immense size of the mineral region, the exceedingly superficial examination, the vast area yet unexplored, and the absence of scientific knowledge on the part of those whose discoveries were only prompted by the unusual appearance of minerals, have led to overlooking many that are of value to the scientist and commerce.

Many minerals given are common to several localities in the State, but, as observed, non-exploration has prevented a complete knowledge of their locations. To make subsequent editions of this work more complete, the compiler solicits the aid of the reader. Any information regarding localities of minerals will be greatly appreciated, and where the variety of a mineral is in doubt, specimens sent will be investigated.

The compiler is under special obligations to Mr. WHITMAN CROSS and Mr. W. B. SMITH, of the U. S. Geological Survey, for valuable information.

The following are titles of the works referred to in the following pages :

Proceedings of the Colorado Scientific Society.
Proceedings of the Academy of Natural Science, Philadelphia.
Proceedings of the American Philosophical Society, Philadelphia.
American Journal of Science.
American Chemical Journal.
Bulletins of the U. S. Geological Survey.

J. S. RANDALL.

GEORGETOWN, COLORADO,
January, 1887.

Minerals of Colorado.

GOLD.

Crystallizes in the Isometric system; the octahedron and dodecahedron are the most common forms; sometimes in cubes and trapezohedrons; the crystals are usually small and imperfect, and sometimes elongated; also reticulated, arborescent and spongiform shapes, in scales and rolled masses. Color various shades of gold-yellow, sometimes inclining to silver-white. Very ductile and malleable. Hardness 2·5 to 3. Specific gravity 19·34 when quite pure.

Occurs in greater or less quantities throughout the mountainous portions, of the State in the gravel of streams and in veins. The veins of Gilpin County are especially noteworthy for the amount of gold mined annually. Small but unique crystals have been found in a number of the mines. Large masses of wire and leaf gold were taken from the Printer Boy mine, in Lake County, in 1870. In combination with tellurium in many of the mines in Boulder County.

SILVER.

Isometric; occurs crystallized in cubes, octahedrons, trapezohedrons and other forms, but generally distorted; usually filiform, arborescent, in plates or coatings and sometimes massive. Color silver-white, but subject to tarnish. Ductile. Hardness 2·5–3. Gravity 10·5 when pure.

Found native in small quantities in nearly all silver mines in the State. In scales, arborescent and filiform, the threads sometimes knotting a mass two inches in diameter, in the mines of Georgetown. Occasionally found in nuggets from the size of a pea to those weighing a pound. The Boulder Nest mine in Clear Creek County, and the Caribou mine in Boulder County were especially productive of native silver in their surface workings. Frequently found in the Blue River placers. Crystals are very seldom met with.

MERCURY. Quicksilver.

Isometric; Occurs in small fluid globules. Color tin-white. Lustre metallic.

Has been found sparingly in the tellurium-bearing mines of Boulder County.

NATIVE COPPER.

Isometric; occurs crystallized in numerous and often compound forms; often filiform and arborescent; also massive. Color copper-red. Streak metallic shining. Ductile and malleable. H. 2·5–3. G. 8·94. Usually found alloyed with some copper, bismuth, etc.

At the head of Hand-Cart Gulch, where it was mined in early days under the supposition that it was gold. Occurs in small quantities in the mines of Central City. Quite large quantities of wire copper have been found in the Pittsburg mine at Empire. Dendritic coatings found in the Centennial mine, Georgetown. Loose masses have been found in the soil near Idaho Springs. Quite abundant in the copper lodes near Golden, where it occurs in seams from $\frac{1}{4}$ to one inch in thickness.

NATIVE BISMUTH.

Hexagonal; occurs in rhombohedrons nearly approaching the cube; also reticulated and arborescent shapes; foliated and granular. Color silver-white with a reddish tinge, but subject to tarnish. Cleavable. Sectile. Brittle when cold, but somewhat malleable when heated. H. 2–2·5. G. 9·72.

Small nuggets of native bismuth are frequently found in the placers of French Gulch, Summit county. Occasionally found in the Las Animas mine, Boulder county, associated with bismutite and bismuthinite. Specimens in collection of J. S. Randall. Bismuth is used extensively in the arts and in medicine; with antimony in the construction of the most sensitive thermometers; with mercury for silvering glass; the sub-nitrate is used for enameling porcelain, and in gilding; in the manufacture of porcelain and optical glass; the nitrate is used in dyeing lilac and violet in calico printing.

NATIVE TELLURIUM.

Hexagonal; occurs in six-sided prisms with the terminal edges replaced by single planes; generally massive and granular. Color tin-white. Lustre metallic. H. 2–2·5. G. 6·1–6·3. Usually carries some gold.

Found in the mines of Boulder county, associated with other tellurium minerals. A few crystals have been found nearly an inch long. Specimens in compiler's collection. Tellurium has no use in the arts, and is valuable only for its gold and silver contents.

NATIVE SULPHUR.

Orthorhombic; the prevailing form of crystal is an acute octahedron, composed of two four-sided pyramids with rhombic bases; usually occurs in masses, sometimes consisting of concentric coats. Color sulphur-yellow, sometimes reddish or brown. Lustre resinous. Transparent to subtranslucent. H. 1·5–2·5. G. 2·072.

Abundant about the mud volcanoes in the southwestern part of the State.

JAROSITE. Sulphate of Iron. *G. A. Konig, Proc. Acad. Nat. Sci., Phila.*
Rhombohedral; occurs in crystals; also massive granular, fibrous, as nodules and as an incrustation. II. 2·5. G. 3·5. Color ochre-yellow, amber-yellow to dark brown. Lustre resinous to adamantine. Analysis by Konig of mineral from Iron Arrow mine: sulphuric acid 28·57, sesquioxyd of iron 51·10, water 10·26, potassa 7·13, and a little soda and silica.

Found in minute brilliant crystals in the Iron Arrow mine, Chaffee county.

ANGLESITE. Sulphate of Lead.
Orthorhombic; occurs in rhombic prisms which assume the general form of the octahedron when short; often in slender implanted crystals; also occurs massive, lamellar or granular. Color gray, white, yellowish or slightly green. Lustre adamantine, inclining to resinous. Transparent to opaque. II. 3. G. 6. Comp., oxide of lead 73·6, sulphuric acid 26·4.

In the Freeland mine, Clear Creek county, in aggregations of grayish translucent crystals about the size of a grain of wheat. In many of the Leadville mines in small crystals and in crystalline masses. A product of the decomposition of galena.

NICCOLITE. Arsenical Nickel.
Hexagonal; rarely occurs crystallized; usually massive, sometimes reniform with a columnar structure. H. 5–5·5. G. 7·33–7·67. Lustre metallic. Color pale copper-red, with a gray or black tarnish. Streak pale brownish-black. Comp., arsenic 55·9, nickel 44·1.

Gem mine, on Pine Creek, Fremont county, associated with bornite and pyritiferous minerals, in dolomite gangue. Produces annabergite on alteration (*Cross*). In the mines at Silver Cliff (*Genth*). Sparingly in the Rosa mine, at Dumont, Clear Creek county.

MELONITE. Telluride of Nickel. *W. F. Hillebrand, Proc. Col. Sc. Soc.*
Hexagonal; generally occurs in indistinct granular and foliated particles. Lustre metallic. Color reddish-white, rarely tarnished brown. Comp., tellurium 76·49, nickel 23·51.

Occurs in the Forlorn Hope mine, Boulder county, in crystalline forms, associated with other tellurides. Specimens in collection of Colorado Scientific Society. A rare mineral, heretofore found only in California.

PHENACITE. Silicate of Glucina. *W. Cross, Am. Jr. Sc., Oct.,* 1882.
Rhombohedral; crystals occasionally oblong, but generally obtuse. Colorless or bright wine-yellow inclining to red; lustre vitreous; transparent to opaque; double-refracting. H. 7·5–8. G. 3. Comp., silica 54·2, glucina 45·8. Resembles quartz.

At Crystal Park, El Paso county, in cavities in granite, with topaz, zircon, microcline, albite, smoky quartz and limonite in large, broken crystals. At Florissant in small, flat crystals upon amazonstone with albite, gothite and smoky quartz. Specimens in collection of Whitman Cross, Denver.

SMALTITE. Arsenide of Cobalt. *W. M. Iles, Am. Jr. Sc.*, May, 1882.
Isometric; occurs in octahedrons, cubes and their modifications; also massive and in imitative shapes. Color tin-white, inclining to steel-gray when massive, sometimes iridescent. Lustre metallic. H. 5·5–6. G· 6·4 to 7·2. Analysis of Colorado mineral by Iles: cobalt 11·59, iron 11·99, arsenic 63·82, with traces of lead, bismuth and sulphur.

Abundant in veins near Gothic, Gunnison County, associated with erythrite, native silver, pyrite and calcite. Produces erythrite by alteration.

ALASKAITE. (*new*). Sulphide of Bismuth, Lead and Silver. *G. A. Konig, Am. Phil. Soc. Phil.*, 1881.
Massive, small foliated, with occasional cleavage planes. Color whitish lead-gray. Lustre metallic. Easily pulverized. G. 6·878. Comp., sulphur 17·63, bismuth 56·97, lead 11·79, silver 8.74, copper 3·46, antimony 0·62, zinc 0·79.

Occurs intimately mixed with quartz, barite, chalcopyrite and tetrahedrite at the Alaska mine, Poughkeepsie Gulch. San Juan county. Named after the Alaska mine.

BEEGERITE. (*new*.) Sulphide of Lead, Copper and Bismuth. *G. A. Konig, Am. Chem. Jr.*, 1881.
Isometric; in elongated crystals; also massive. Cubic cleavage. Color light to dark gray. Lustre brilliant metallic. Comp., sulphur 14·97, bismuth 20·59, lead 64·23, copper 1·70.

From the Baltic lode, near Grant, Park county. Named after Mr. Herman Beeger, of Denver.

TENNANTITE. Arsenical Sulphide of Copper and Iron.
Isometric; usually occurs in rhombic dodecahedrons, sometimes variously modified; also in cubes and octahedrons, of which the edges and angles are replaced. Color lead-gray to iron-black. Lustre metallic. H. 3·5–4. G. 4·37. Comp., varies, copper 48, sulphur 30, arsenic 13, iron 10.

Occurs in the Freeland mine, Clear Creek county, in clusters of splendent crystals from ⅓ to ½ an inch in diameter. A few dodecahedrons are met with, but most of the crystals appear complicated. Fine specimens in collection of compiler. This is the first known occurrence of the mineral in the United States.

ZINKENITE. Antimonial Sulphide of Lead. *W. F. Hillebrand, Proc. Col. Sci. Soc., vol. 1.*
Orthorhombic; crystals regular six-sided prisms terminated by low six-sided pyramids; faces of prisms generally deeply striated longitudinally; also massive, fibrous and columnar. Resembles stibnite, but is harder and heavier. Color steel-gray. Lustre metallic. H. 3–3·5. G. 5·30–5.35. Analysis of Red Mountain mineral by Hillebrand: antimony 35·00, arsenic 5·64, lead 32·77, copper 1·22, silver 0·23, sulphur 22·50, and small amounts of iron, soda, etc.

Occurs massive with apparent crystalline structure, of iron-gray color, in the Brobdignag mine, Red Mountain, San Juan county. This is the first observed occurrence of the mineral in the United States.

CHALCOPYRITE. Copper Pyrites. Sulphuret of Copper and Iron.
Tetragonal; the crystals present the general form of the tetrahedron but
are sometimes compound; usually occurs massive and frequently amorphous. Color brass-yellow, often with a variegated taruish. Lustre metallic. H. 3·5–4. G. 4·15. Comp., sulphur 34·9, copper 34·6, iron 30·5.

Massive, in greater or less quantity, in nearly all mines in
the state, and nearly always highly auriferous when found in
gold mines, and argentiferous when found in silver mines.
That from the mines of Red Elephant Mountain, Clear Creek
county, usually carries several hundred ounces of silver in
a ton of ore. Fine tetrahedrons, measuring from ¼ to ½ an
inch are found in the Freeland mine, Clear Creek county, but
nearly all are incrusted with minute crystals of tennantite.
Fine specimens in collection of J. S. Randall.

BISMUTHINITE. Bismuth Glance. Sulphide of Bismuth.
Orthorhombic; occcurs in acicular prisms, striated longitudinally; also
massive with a foliated or fibrous structure. Color lead-gray inclining to
tin-white, with a yellowish or iridescent tarnish. Lustre metallic. H. 2.
G. 6·4–7·16. Comp., sulphur 18·75, bismuth 81·25.

In working a quartz vein near Guy Hill, Jefferson county,
a single "pocket" of bismuthinite was encountered, and several hundred pounds of the ore taken out. Occurs in stout
columnar aggregations in the Las Animas mine, near Gold
Hill, Boulder county. In quartz veins on Big Thompson
Creek, Larimer county. In small crystals and masses, with
a gold-yellow tarnish, in the Little Giant lode, Clear Creek
county. Beautiful specimens in collection of compiler.

LOLLINGITE. Arsenide of Iron. *W. F. Hillebrand, Am. Jr. Sc., May,*
1884; *Proc. Col. Sc. Soc., vol. 1.*
Orthorhombic; occurs in right rhombic prisms, parallel to whose planes it
may be cleaved; also massive, acicular and columnar. H. 5–5·5. G. 6·2,
7·43. Color tin-white. Lustre metallic. Analysis of Colorado mineral
by Hillebrand: arsenic 71·18, iron 22·96, cobalt 4·37, sulphur 0·56, bismuth 0·08, copper 0·39, nickel 0·21.

In the mines on Teocalli and White Rock Mountains, in
Gunnison county, in dense radiate masses, and more rarely in
trillings and minute crystals, associated with smaltite, marcasite, galena, chalcopyrite, pyrargyrite, argentite, proustite
and native silver, in a gangue of calcite, siderite and barite.
Specimens in collection of Colorado Scientific Society.

GUITERMANITE. *(new)*. Arsenical Sulphide of Lead. *W. F. Hille-*
brand, Proc. Col. Sc. Soc., vol. 1.
Occurs massive of a bluish-gray color and slight metallic lustre. H. 3.
G. 5·94. Analysis by Hillebrand: arsenic 13·00, sulphur 19·56, lead 61·63,
copper 0·17, silver 0·02, iron 0·88, oxygen 0·55.

Occurs massive in the Zuni mine, near Silverton, and forms
the matrix of zunyite. Named after Franklin Guiterman,
who first brought the mineral to notice.

RHODOCHROSITE. Carbonate of Manganese.
Rhombohedral; occurs in saddle-shaped lenticular crystals; also botryoidal and massive. H. 3·5–4·5. G. 3·4–3·7. Listre vitreous inclining to pearly. Color shades of rose-red, dark red, brown, gray. Translucent. Structure lamellar. Comp., carbonic acid 38·6, protoxyd of manganese 61·4.

In beautiful rose-red translucent single crystals in the Royal Albert and Mountain Monarch veins, Uncompahgre district, Ouray county. In large opaque clusters in the Champion vein, Burrough's Park, Hinsdale county (*Cross*). In small white radiate bunches upon blue celestite, in the Garden of the Gods (*Hillebrand*). Some fine clusters of crystals have been found in the Danube mine, near Idaho Springs, Clear Creek county. Specimens in collection of J. S. Randall.

BROCHANTITE. Hydrous Sulphate of Copper. *Richard Pearce, Proc. Col. Sc. Soc., vol.* 1.
Orthorhombic; Occurs in well-defined tabular and acicular crystals and drusy crusts; crystals vertically striated; also massive, and reniform with a columnar structure. H. 3·5–4. G. 3·78. Color emerald-green to blackish-green. Lustre vitreous. Transparent, translucent. Comp., sulphuric acid 19·9, protoxide of copper 69, water 11·1.

Occurs in considerable quantity and very pure, with partial development of crystal form, in the Monarch mine, Chaffee county.

FLUORITE. Fluor Spar. Fluorid of Calcium.
Isometric; occurs in cubes, octahedrons, rhombic dodecahedrons and their modifications; crystals often have their faces made up of small cubes, or cavernous with rectangular cavities; also massive, coarse or fine granular. Easily cleaved into the tetrahedron, acute rhombohedron and octahedron. H. 4. G. 3·1–3·25. Color white, grey, and various tints of blue, green, yellow, purple and red. Transparent to opaque. Lustre vitreous, sometimes splendent. Fragments heated or rubbed against each other in the dark become luminous. Comp., fluorine 48·7, calcium 51·3.

Large veins of quite pure purple fluorite have been found on St. Vrains Creek and Jim Creek, Boulder county, but are undeveloped, there being no demand for it at present. On Bear and Cub Creeks, Jefferson county, associated with galena. Clusters of transparent crystals, of green color, are found on Plum Creek, Douglas county. Some very good crystallized specimens are found in the Crystal Mountains, El Paso county, associated with amazonstone, gothite, etc. Massive and crystallized of a deep purple color on Tarryall Creek, South Park. It also occurs sparingly in many of the mines throughout the state. Fluor spar is used as a flux in the reduction of ores, and is the source of hydrofluoric acid for etching glass. In England it is wrought into vases and other articles for ornament. Specimens in collection of J. S. Randall.

CRYOLITE. Fluoride of Sodium and Aluminum. *Cross and Hillebrand, Am. Jr. Sc., Oct.,* 1883.

Orthorhombic; crystals resemble the cube; usually occurs massive with a rectangular cleavage. Color usually snow-white, sometimes brown or black; lustre vitreous; translucent but becomes more transparent when immersed in water. Melts in the flame of a candle. H. 2·5. G. 3. Analysis of Colorado mineral by Hillebrand: Fluorine 53·55, sodium 32·40, aluminum 12·90, iron oxide, calcium and water.

In small masses in quartz and feldspar veins, with astrophyllite, zircon and columbite, near the toll road at St. Peter's Dome, west of Cheyenne Mountain, El Paso county. Specimens of the cryolite groupe of minerals—cryolite, pachnolite, prosopite, thomsenolite and gearksutite—in collection of Colorado Scientific Society. Specimens of cryolite in collection of J. S. Randall. Cryolite is used for making soda and soda and alumina salts, and white glass. Produces pachnolite on alteration.

PACHNOLITE. Fluoride of Aluminum, Calcium and Sodium. *Cross and Hillebrand, Am. Jr. Sc., Oct.,* 1883.

Monoclinic; crystals usually four-sided prisms; also occurs massive. Colorless to white or yellowish. Transparent to translucent. Analysis by Hillebrand of Colorado mineral: aluminum 12·36, calcium 18·04, sodium 10·25, water 8·05, fluorine 51·30.

Massive and in small transparent crystals, at St. Peter's Dome, El Paso county. Product of the alteration of cryolite. Specimens in collection of compiler.

PROSOPITE. Fluoride of Aluminum and Calcium. *Cross and Hillebrand, Am. Jr. Sc., Oct.,* 1883.

Monoclinic; occurs in minute imbedded crystals. Colorless, white or grayish. H. 4·5. G. 2·89. Analysis of Colorado mineral by Hillebrand, fluorine 35·01, aluminum 23·37, calcium 16·19, magnesium 0·11, sodium 0·33, water 12·41, loss as oxygen 12·58.

At St. Peter's Dome, massive and in small crystals, with other fluorides.

THOMSENOLITE. Fluoride of Aluminum, Calcium and Sodium. *Cross and Hillebrand, Am. Jr. Sc., Oct.,* 1883.

Monoclinic; occurs in slender prisms, horizontally striated ; also massive, opal or chalcedony-like. Color white, sometimes reddish. Lustre vitreous to waxy. Transparent to translucent. H. 2·5-4. G. 2·75. Comp., fluorine 52·2, aluminum 15·0, calcium 15·4, sodium 7·6, water 9·8.

Found sparingly, mixed with pachnolite, at St. Peter's Dome.

GEARKSUTITE. Fluoride of Aluminum, Calcium and Sodium. *Cross and Hillebrand, Am. Jr. Sc., Oct.,* 1884.

Occurs in minute needles and kaolin-like. Color white. Lustre dull. H. 2. Analysis of Colorado mineral by Hillebrand: aluminum 15·31, calcium 22·30, fluorine 42·07, water 15·46, sodium 0·10, potassa 0·04.

Occurs quite abundant as minute crystals and kaolin-like, at St. Peter's Dome.

ORTHOCLASE. Potash Feldspar.

Monoclinic; occurs in modified oblique rhombic prisms; usually thick often rectangular, and also in modified tables; also massive, sometimes lamellar. Color white, gray and flesh-red common; also greenish and bluish, when it is called Amazonstone. Lustre vitreous, sometimes pearly on a cleavage face. Cleaves easily into plates, Transparent to subtranslucent. H. 6. G. 2·39–2·62. Comp., silica 64·20, alumina 18·40, potash 16·95.

The common feldspar of granitic rocks. Fine crystals are abundant in the Lida the Little Queen lode at Kokomo. Specimens in collection of J. S. Randall.

MICROCLINE. Feldspar. Variety Orthoclase.

Color pink and gray. Composition, etc., same as orthoclase.

Associated with amazonstone, etc., near Fiorissant, and occurring in the same crystalline forms.

AMAZONSTONE. Green Feldspar. Variety Microcline.

Crystallizes in the same system, and has nearly the same composition as orthoclase. Color bluish-green. According to Konig, the the color is due to iron.

Fine specimens near Florissant, El Paso county, in cavities in granite, associated with smoky quartz, albite, topaz, phenacite, fluorite, gothite and columbite. The crystals are often of great size and frequently in groups. Baveno and Carlsbad twin crystals are quite frequent. Fine specimens in the compiler's cabinet.

ALBITE. Soda Feldspar.

Triclinic; occurs generally in flat twin crystals, which are often reversed, the one upon the other; also massive, either lamellar or granular, the laminæ sometimes divergent; granular varieties often quite fine. Color varies from milk-white to green or red. Lustre pearly upon a cleavage face. Transparent to opaque. H. 6–7. G. 2·59. Comp., silica 68·6, soda 11·8, alumina 19·6.

Near Florissant with amazonstone, sometimes capping crystals of that mineral, but more frequently forming the base or matrix in which the crystals are set. Often in granite rocks.

PITCHSTONE. Volcanic Glass. A fused feldspar.

Occurs massive. Color tints of green, yellow, red, brown and black. Lustre pitch-like or glassy. Translucent to opaque. Breaks with a sharp edge. H. 5–6. Comp., either orthoclase, albite or oligoclase.

Forms a dyke on the north wall of the Colorado Central mine, Georgetown. Found near Del Norte.

ZUNYITE. (*new.*) Silicate of Alumina. *W. F. Hillebrand, Proc. Col. Sci. Soc., vol. 1.*

Isometric; occurs in tetrahedral prisms with similar terminations. Colorless when pure, but often black from impurities. Lustre vitreous. Transparent to opaque. H. 7. G. 2·87. Analysis by Hillebrand: silica silica 24·33, alumina 57·88, lithia 10·89, fluorine 5·61, chlorine 2·91, with traces of iron, soda, etc.

Occurs in very small crystals imbedded in guitermanite from the Zuni mine near Silverton. Named after that mine.

COSALITE. Sulphide of Lead and Bismuth. *W. F. Hillebrand, Am. Jr., Sc., May,* 1884; *Proc. Col. Sc. Soc., vol.* 1.
Crystal apparently rhombic, longitudinally striated. Color lead-gray, grayish-white, pale yellow on exposed surface. Lustre metallic. H. 3·5.
Analysii of mineral from Comstock mine, by Hillebrand : bismuth 42·93. lead 22·49, sulphur 17·11, silver 8·43, copper 7·50.
Occurs in irregular masses, rarely an inch in length, without crystalline structure, in the Comstock mine, near Parrott City, La Plata county, associated with pyrite, sphalerite, a telluride and gold in a quartz vein. Massive in the Yankee Girl mine, San Juan county. Specimens in collection of Colorado Scientific Society.

AUTUNITE. Lime-Uranite. Phosphate of Uranium and Lime.
Orthorhombic; crystals very nearly square and cleavable. Color citron- to sulphur-yellow. Streak yellowish. Translucent. Lustre pearly to adamantine. H. 2–2·5. G. 3·05–3·19. Comp., phosphoric acid 15·7, oxyd of uranium 62·7, lime 6·1, water 15·5.
Found in the Peabody lode, near Georgetown, in minute greenish-yellow, tabular crystals. Also in a lode on Chicago Creek, Clear Creek county.

ENARGITE. Arsenical Sulphide of Copper. *B. S. Burton, Am. Jr. Sc., Jan.,* 1868.
Orthorhombic. Massive, granular or columnar. Cleavage perfect giving a brilliant metallic luster. Color grayish to iron-black. Streak grayish-black, powder having a metallic luster. Brittle. H. 3. G. 4·43. Analysis of Willis Gulch mineral by Burton : sulphur 31·56, copper 47·58, arsenic 17·80, antimony 1·37, iron 1·04.
Occurs massive and crystallized in the Powers lode, Willis Gulch, Gilpin county. Specimens in the compiler's collection.

ACANTHITE. Silver Glance under an orthorhombic form.
Orthorhombic; crystals usually slender-pointed prisms. Luster metallic. Color iron-black. Sectile, cutting like lead. Fracture uneven, giving a shining surface. H. 2·5 or under. G. 7·16–7·33. Comp., sulphur 12·9, silver 87·1.
Occurs in small crystals in the Little Emma mine, Georgetown. Specimens in compiler's collection.

ARGENTITE. Silver Glance. Vitreous Silver. Sulphuret of Silver.
Isometric; primary form the cube; also occurs in octahedrons and rhombic dodecahedrons; reticulated, dendritic, stalactitic, amorphous and massive. Color blackish lead-gray. Streak shining. Luster metallic. Fracture small sub-conchoidal, uneven. Perfectly sectile, cutting like lead. H. 2–2·5. G. 7·196–7·365. Comp., sulphur 12·9, silver 87·1.
This valuable ore of silver occurs in small quantities in nearly all the silver mines in the state, but is seldom found in masses weighing more than two or three pounds, although frequently found in sufficient quantities to make the ore with which it is associated a high grade silver ore. Very good crystallized specimens are occasionally met with in the Georgetown mines. Specimens in compiler's collection.

SILVER BLACK. Sulphuret of silver in the form of a powder or stain.
Of frequent occurrence at the surface of silver-bearing
veins. Commonly called "sulphurets."

CHALCEDONY. Cryptocrystalline variety of Quartz.
Occurs in mammillated, botryoidal and stalactitic forms, but never in a
crystallized state. Luster like that of wax. Color white, grayish, pale-
brown to dark-brown and black, rarely delicate blue. H. 7. G. 2·5–2·8.
Comp., silica, with some opal silica.

Abundant on the ridge to the left of Willow Creek in Middle Park; also in the basalt on Corral Creek. Handsome
pale blue specimens near the Salt Works in South Park; six
miles south of the Salt Works in jasper and semi-opal. Fine
specimens in San Luis Park on La Garita, filling cavities of
amygdaloid. At Los Pinos Agency, in basalt. At the head
of Cherry Creek, pseudomorphous after wood. On the Rio
Grande, 15 miles above Loma, large geodes. Blue specimens
on the Gunnison above the Grand Canon. Common in the
valley of the Gunnison near Grand River. Fine specimens in
the compiler's collection. *

BORNITE. Erubescite. Purple Copper Ore.
Isometric; the crystals are generally cubes of which the solid angles are
replaced, and the faces are mostly curvilinear. Usually massive. Color
between copper-red and pinchbeck-brown, but it soon acquires an iridescent tarnish. Streak pale grayish-black. Luster metallic. Brittle. H.
3. G. 4·4–5·5. Comp., varies, sulphur 22·11, copper 70·13, iron 7·76.

A valuable gold-ore in the gold mines, and in some mines
is quite abundant. Frequently found in silver-bearing veins,
and sometimes becomes a valuable ore of silver. In the compiler's collection are some very good crystals from Central.

CHABAZITE. *Cross & Hillebrand, Am. Jr. Sc., June,* 1882; *Bulletin No.*
20, U. S. Geo. Sur.
Rhombohedral; usually occurs crystallized in obtuse rhombohedrons, and
has a rhombohedral cleavage. Color white and reddish. Luster vitreous. Transparent to translucent. H. 4–5. G. 2·08–2·19. Analysis of
Golden mineral by Hillebrand: silica 47·18, alumina 19·67, potash 0·37,
lime 9·74, strontia 9·43, soda 0·51, water 25·15.

Good glassy-white specimens are quite abundant in the
amygdaloid of North Table Mountain, Golden. Specimens
of all the Table Mountain minerals in the compiler's cabinet.

STILBITE. Radiated Zeolite. *Bulletin No. 20, U. S. Geo. Sur.*
Orthorhombic; usually occurs in prisms of which the edges are replaced,
and with four-sided summits; often in sheaf-like aggregations and in diverging groups; also massive, in radiating and broad columnar forms.
Color white, sometimes yellow, gray, red or brown. Transparent to translucent. Luster vitreous. Double refracting. H. 3·5–4. G. 2·094–2·161.
Analysis of Golden mineral by Hillebrand: silica 54·67, alumina 16·78,
lime 7·98, soda 1·47, water 19·16.

Occurs sparingly in small, clear crystals in the amygdaloid
of North Table Mountain, Golden.

LAUMONTITE. *Cross and Hillebrand, Am. Jr. Sc., Aug.,* 1882, *U. S. Geo. Sur., Bul.* 20.
Monoclinic; occurs in columnar, radiating and compact crystalline masses, and in separate crystals. Color white or yellowish-white, sometimes reddish. Luster vitreous to earthy. Transparent to opaque; becoming opaque and sometimes pulverulent on exposure. H. 3·5–4. G. 2·25–2·36. Analysis of Table Mountain mineral by Hillebrand: silica 52·07, alumina 21·30, lime 11·24, potassa 0·42, soda 0·48, water 14·58.
Occurs as a reddish-yellow sand-like material, and in compact crystalline masses in the cavities of the amygdaloid of North Table Mountain at Golden.

THOMSONITE. *Cross and Hillebrand, Am. Jr. Sc., June,* 1882; *Bulletin No.* 20, *U. S. Geo. Sur.*
Orthorhombic; occurs in right rectangular prisms, with cleavage parallel to its side; usually massive with a columnar or radiated structure; also in radiated spherical concretions. Luster vitreous inclining to pearly. Colorless or snow-white; impure varieties brown. Transparent to translucent. Pyroelectric. H. 5–5·5. G. 2·3–2·4. Analysis of Golden mineral by Hillebrand: silica 42·66, alumina 29·25, lime 10·90, soda 4·92, water 12·23.
Occurs in the amygdaloid of Table Mountains, Golden, in minute rectangular blades, which are placed upon each other like the leaves of a closed fan, and in spherical concretions having a radiated structure.

ANALCITE. *Cross and Hillebrand, Am. Jr. Sc., June,* 1882; *Bulletin No.* 20, *U. S. Geo. Sur.*
Isometric; usually occurs in trapezohedrons, or 24-sided crystals; also massive. Luster vitreous. Transparent to opaque. Colorless, white, occasionally grayish, greenish, yellowish or reddish-white. H. 5–5·5. G. 2·22–2·29. Analysis of Golden mineral by Hillebrand: silica 55·80, alumina 22·45, soda 13·45, water 8·35.
Occurs in pure white or transparent crystals, and vary in size from small ones to those nearly an inch in diameter, in the amygdaloid of North Table Mountain, Golden. On the eastern side of the mountain, analcite is specially abundant. It is also abundant on South Table Mountain, but the crystals are usually quite small.

APOPHYLLTE. *Cross and Hillebrand, Am. Jr. Sc., August,* 1882; *Bulletin No.* 20, *U. S. Geo. Sur.*
Tetragonal; occurs in right square prisms or octahedrons, which often terminate in a sharp pyrimid; crystals sometimes nearly cylindrical or barrel-shaped; also massive and lamellar. Cleavage highly perfect to all the planes of the primary form. Color white or grayish, sometimes greenish, yellowish, or reddish. Luster vitreous to pearly. Transparent to opaque. H. 4·5–5. G. 2·3–2·4. Analysis of Golden mineral by Hillebrand: silica 51·89, alumina 1·54, iron 0·13, lime 24·51, potash 3·81, soda 0·59, water 16·52, fluorine 1·70.
Occurs in well-developed crystals of prismatic habit in the amygdaloid of North Table Mountain, Golden. The larger crystals which are occasionally half an inch in diameter, are often of a greenish tinge, and have uneven surfaces and are terminated by a large number of small pyramids.

A pearly-white, very finely-foliate substance resembling albin is produced by the alteration of apophyllite, which according to Hillebrand, consists of silica 67·96, alumina 8·48, iron oxyd 1·04, lime 5·47, magnesia 0·53, potassa 1·23, soda 0·74, water 14·55.

MESOLITE. Fibrous Zeolite. *Cross and Hillebrand, Am. Jr. Sc., August,* 1882; *Bulletin No. 20, U. S. Geo. Sur.*
Occurs in long slender crystals, often very delicate; also in silky fibrous or columnar masses. Luster of crystals vitreous; of fibrous masses more or less silky. Colorless or white, grayish or yellowish. Fragile. Transparent to translucent. H.5. G. 2·2-2·4. Analysis of Golden mineral by Hillebrand: silica 46·17, alumina 26·88, lime 8·77, soda 6·19, water 12·16.

Occurs in the amygdaloid of North Table Mountain, Golden, in masses composed of exceedingly delicate needles loosely grouped together, and sometimes as a continuous membrane, or like a thick cobweb.

NATROLITE. Needle Zeolite. *Cross and Hillebrand, Bulletin No. 20, U. S. Geo. Sur.*
Orthorhombic; crystals usually slender, often acicular; frequently interlacing, divergent, or stellate; also fibrous, radiating, massive, granular, or compact. Luster vitreous to pearly. Transparent to translucent. Color white or colorless, yellowish, grayish or reddish. H. 5-5·5. G. 2·17-2·25. Analysis of Table Mountain mineral by Hillebrand: silica 43·66, alumina 24·89, lime 4·87, soda 14·66, water 8·09.

Natrolite is the least abundant of the zeolites, and has been found only on the northern part of South Table Mountain, where it appears in delicate prisms sparingly deposited upon analcite or associated with that mineral. It has also been observed upon yellow calcite, chabazite and thomsonite.

LEVYNITE. *Cross and Hillebrand, Bulletin No. 20, U. S. Geo. Sur.*
Rhombohedral; occurs in twin crystals, the faces of which are often striated; frequently in druses. Colorless, white, grayish, greenish, reddish, yellowish. Luster vitreous. Transparent to translucent. H. 4-4·5. G. 2·09-2·16. Analysis of Golden mineral by Hillebrand: silica 46·76, alumina 21·91, lime 11·12, potassa 0·21, soda 1·34, water 18·65.

Small white and colorless crystals occur sparingly in the amygdaloid of North Table Mountain, Golden. Associated with the levynite is a fibrous mineral, dull white in color and never showing crystal faces, which is almost identical with the levynite in composition.

SCOLECITE. Fibrous Zeolite. *Cross and Hillebrand, Bulletin No. 20, U. S. Geo. Sur.*
Monoclinic; occurs in long or short prismatic or acicular crystals; very often in twins; also massive with a fibrous or radiating structure. Colorless, snow-white, grayish yellowish, and reddish. Transparent to translucent. Luster vitreous, or silky when fibrous. H. 5-5·5. G. 2·16-2·40. Analysis of Golden mineral by Hillebrand: silica 46·03, alumina 25·28, iron oxyd 0·27, lime 12·77, soda 1·07, potassa 0·13, water 14·48.

Occurs sparingly in small cavities in a zone just above that containing a great number of zeolites on North Table Mountain, Golden. It appears in small spheres or segments of spheres with a radiate structure, and resembles thomsonite, though easily distinguished by the brilliant white color and satin-like luster.

BOLE. Variety of Halloysite. Hydrous Silicate of Alumina. *Cross and Hillebrand, Bulletin No.* 20, *U. S. Geo. Sur.*
Occurs massive; clay-like or earthy. Luster somewhat pearly or waxy. Feels greasy. Colors usually dark. H. 1–2. G. 1·8–2·4. Analysis of Golden mineral by Hillebrand: silica 46·17, alumina 22·03, iron oxyd 4·64, lime 2·30, magnesia 2·42, potassa and soda 2·06, water 20·38.

Occurs in the amygdaloid about the center of South Table Mountain, Golden, as a dark brown clay. Placed in water it falls apart without decrepitation, unlike ordinary bole. The mineral is used as a pigment.

HUBNERITE. Tungstate of Manganese. *Cross and Hillebrand, Bulletin No.* 20, *U. S. Geo. Sur.; Proc. Col. Sc. Soc., vol.* 1.
Orthorhombic; usually occurs in columnar masses or foliated. Luster adamantine on face of cleavage; elsewhere greasy. Color brownish-red, brownish-black, yellowish. Streak yellowish-brown. Opaque. H. 4·5. G. 7·4. Analysis of mineral from Royal Albert vein by Hillebrand: silica 0·62, tungstic acid 75·58, protoxyd of manganese 23·40, protoxyd of iron 0·24, lime 0·13, columbic acid (?) 0·05.

The mineral occurs in the Royal Albert vein, Uncompahgre district, Ouray county, in long flattened crystals vertically striated, of a brownish-black to pale yellow color, imbedded in quartz. It is also found near Silverton, and at Jimtown, Boulder county, where it appears in small blades disseminated through quartz. Specimens in compiler's cabinet.

HELIOTROPE. Bloodstone. Cryptocrystalline Quartz.
A bright to leek-green variety of chalcedony or jasper, with small spots of red jasper looking like drops of blood.

Occurs in a vein of jasper in the hill at the junction of Willow Creek and the Grand River, Middle Park. On account of its beautiful color, heliotrope has always been much used for rings, seals and other ornaments. Specimens of the different forms of quartz in the compiler's cabinet.

CARNELIAN. Cryptocrystalline Quartz.
A variety of chalcedony (p. 14) of a clear bright red tint, pale to deep in shade. The brownish-red is also called carnelian. The color is due to the presence of iron.

Found sparingly on the ridge to the left of Willow Creek in Middle Park; on the Rio Grande near Loma; near Larkspur; on Cherry, Kiowa and Running Creeks, pseudomorphous after wood. From the high polish of which it is susceptible, and its bright colors, carnelian has always, in both ancient and modern times, been much used for ornaments.

PRASE. Cryptocrystalline Quartz.
A translucent leek-green chalcedony (p. 14).
Occasionally found in San Luis Park.

CHRYSOPRASE. Cryptocrystalline Quartz.
An apple-green chalcedony (p. 14), the color of which is due to the presence of nickel.
Occurs sparingly on La Garita, west of San Luis Park. In geodes in Middle Park.

BANDED AGATE. Fortification Agate. Eye Agate. Clouded Agate.
A variegated chalcedony. The colors are either banded or in clouds. The bands are delicate parallel lines of various shades of color. They follow waving or zigzag courses, and are occasionally concentric circular, when it it is called eye agate. When the colors are irregular it is called clouded agate.
Banded and clouded agates are occasionally found on Willow Creek in Middle Park; near the Salt Works in South Park; near the source of Cherry Creek; ten miles south of Canon City, in sandstone; on the La Garita, south of San Luis Park; large geodes on the Rio Grande, 15 miles above Loma. Owing to its beauty, variety, hardness and capability of receiving a high polish, agate is much used both in articles of utility and ornament.

MOSS AGATE. Mocha Stone.
A chalcedony filled with brown, black, green or red moss-like or dendritic forms.
Abundant on Williams Fork, Middle Park; in South Park between Fairplay and the Salt Works.

AGATIZED WOOD.
Wood petrified with clouded agate.
Middle Park near Hot Spring, and on the right of Willow Creek; around the Salt Works in South Park; abundant near the sources of Cherry, Kiowa and Bijou Creeks.

ONYX.
Like agate in consisting of layers, but the layers are in even planes and usually thicker. The colors are generally of a light brown and an opaque white.
Occurs sparingly on the ridge to the left of Willow Creek, in Middle Park, and also near Grand Lake; on the La Garita, west of San Luis Park; a variety called chalcedonyx is found in San Luis Park. Eextensively used for cameos and other articles of adornment.

SARDONYX.
Those varieties of onyx which are composed of alternate layers of red and white.
A few specimens have been found in the Willow Creek region, Middle Park. It is the most beautiful, the rarest and most valuable form of onyx.

PYRARGYRITE. Ruby Silver. Dark Red Silver Ore.
Rhombohedral; crystals usually prismatic; generally occurs in small masses. Lustre metallic-adamantine. Color black, approaching cochineal-red. Streak cochineal-red. Translucent-opaque. H. 2–2·5. G. 5·7. Comp., sulphur 17·7, antimony 22·5, silver 59·8.

Occurs in small masses in most of the silver mines in the state, but rarely found crystallized. Specimens in compiler's collection.

PROUSTITE. Light Ruby Silver Ore.
Rhombohedral; occurs in granular masses and in small crystals. Lustre adamantine. Color cochineal-red. Streak cochineal-red, sometimes inclined to aurora-red. Subtransparent—subtranslucent. H. 2.–2·5. G. 5·42–5·56. Comp., sulphur 19·4, arsenic 15·2, silver 65·4.

Occurs in the mines of Geneva District and about Montezuma in small masses and in clusters of minute crystals. In the Colorado Central mine, Georgetown, in beautiful crystals and massive, associated with galena, sphalerite and crystals of polybasite. Compiler's collection.

COVELLITE. Indigo Copper Ore.
Hexagonal, but rarely occurs in crystals; commonly massive or spheroidal. Color indigo-blue or darker. Streak lead-gray to black, shining. Lustre of crystals submetallic; dull when massive. H. 1·5–2. G. 4·59. Comp., sulphur 33·5, copper 66·5.

Occurs in small grains in the Pewabic lode, Central City. In the form of a powder in some of the mines of Cascade and Daily Districts, Clear Creek County.

COLUMBITE. Columbate of Iron. Ore of Columbium. *J. L. Smith, Am. Jour. Sc.,* May, 1877.
Orthorhombic; occurs in single crystals, crystalline masses, rarely massive. Lustre submetallic; a little shining. Color iron-black, brownish-black, grayish-black; often iridescent. Streak dark red to black. Fracture subconchoidal, uneven. H. 6. G. 5·4–6·5. Analysis, Colorado mineral, by J. L. Smith: columbic acid 79·61, iron protoxide 14·14, manganese protoxide 4·61, loss by heat ·50.

In the Pike's Peak region, in small black acicular crystals imbedded in amazonstone. A few large single crystals and fragments have been also been found. Compiler's collection.

RALSTONITE. *Cross and Hillebrand, Bul. No. 20, U. S. Geo. Sur.*
Isometric; in cubes and octahedrons. Colorless. Transparent. H. 4·5. G. 2·4. Comp., alumina 22·94, fluorine 50·05, calcium 1·99, magnesia 5·52, sodium 4·66, water 14·84.

Occurs in minute crystals in cavities in thomsenolite and pachnolite, at St. Peter's Dome, El Paso County. The crystals here found are transparent cubes whose corners are replaced by small octahedron faces and seldom reach a diameter of 1ᴹᴹ. Some crystals of pachnolite seem to be coated by a crystalline dust whose particles are found under the microscope to be most perfect little crystals of ralstonite.

ELPASOLITE. (*new.*) *Cross and Hillebrand, Bul. No. 20, U. S. Geo. Sur: Am. Jr. Sc.,* October, 1883.

Probably isometric; crystals faces cube and octahedron. Colorless, but not perfectly clear. Calculated analysis by Hillebrand: alumina 11·32, calcium 0·72, magnesia 0·22, potassium 28·94, sodium 9·90, fluorine 46·98.

Sparingly as a compact irregular mass in cavities in massive pachnolite from St. Peter's Dome, El Paso County, from which it was named.

ZIRCON. *Cross and Hillebrand, Am. J. Sc.,* Oct., 1882; *Konig, Am. Phi. Soc.* xvi, 518, 1877.

Tetragonal; primary form an obtuse octahedron with a square base; also occurs in irregular forms and grains. Lustre adamantine. Colorless, pale yellowish, grayish, yellowish-green, brownish-yellow, reddish-brown. Transparent to opaque. Fracture conchoidal and brilliant. Double refraction strong, positive. H. 7·5. G. 4·5–4·75. Comp., silica 33, zirconia 67. Konig made an analysis of zircon from the Pike's Peak region, which was published in the Proceedings of the American Philosophical Society of Philadelphia, in 1877.

Found in numerous localities in the Pike's Peak region, associated with astrophyllite, amazonstone, etc. Near the Pike's Peak toll-road, about due west from Cheyenne Mountain, a prospect tunnel, in following a vein-like mass of white quartz in granite, disclosed zircon imbedded in the quartz so abundant that a cubic inch of the latter mineral contains from 25 to 100 crystals and particles of zircon, varying in size from 1ᶜᵐ downward. Some of these crystals are very perfect, but exhibit only the pyramid, the prisim entirely lacking. Many of them are beautifully clear and of deep reddish-brown, pink, pale honey-yellow and occasionally deep emerald-green color. The perfection of these crystals, with their transparency and color, make them among the most beautiful known. Small irregularly shaped zircons and occacionally elongated crystals occur in the sands of Bear River. They are of various shades of brown, red and yellow, and many are colorless. Most of them are transparent and quite brilliant, but too small to be cut for setting. Specimens from both the above localities in the compiler's cabinet. Although the zircon is but rarely used in jewelry, it makes a beautiful gem when of good color and has a peculiar opalescent reflection.

PICKERINGITE. Magnesia Alum. *Goldsmith, Proc. Acad. Nat. Sci. Phila.,* 1876.

In fine acicular crystals; long fibrous masses; and efflorescences. Lustre silky. Color white or yellowish. Becomes pulverulent and white on exposure. Taste bitter—astringent. H. 1. Analysis of Monument Park mineral by Goldsmith: sulphuric acid 38·69, alumina 11·90, magnesia 4·89, potassa and soda 0·68, sand 1·90, water, by difference, 41·94.

Occurs crystallized in thin needles in the region of Monument Park.

ALTAITE. Telluride of Lead. *Genth, Am. Phi. Soc.,* 1874.
Isometric; usually massive; rarely in cubes. Cleavage cubic. · Lustre
metallic. Color tin-white, resembling that of native antimony, with a yellow tarnish. Sectile. H. 3–3·5. G. 8·159. Analysis by Genth of mineral
from the Red Cloud mine, Boulder county: quartz 0·19, gold 0·19, silver
0·62, copper 0·06, lead 60·22, zinc 9·15, iron 0·48, tellurium 37·99.
Occurs in a number of the tellurium-bearing mines at Gold
Hill, Boulder County. Fine crystals in the Slide mine.

CASSITERITE. Tin-Stone. Oxyd of Tin. *Cross and Hillebrand, Bul.*
20, *U. S. Geo. Sur.*
Tetragonal; primary form an obtuse pyramid with a square base; it is
found in quadrangular prisms, terminated by four-sided pyramids, and in
many more complex forms; often in reniform shapes. Structure fibrous
divergent; also massive, granular or impalpable. Lustre adamantine, and
crystals usually splendent. Color brown or black; sometimes red, gray,
white or yellow. Streak white, grayish, brownish. Nearly transparent to
opaque. H. 6–7. G. 6·4–7·1. Comp., tin 78-67, oxygen 21·33.
Occurs in small crystalline masses imbedded in albite and
quartz on Devil's Head Mountain, Douglas County, associated
with topaz, amazonstone, smoky quartz and fluor spar.
With amazonstone at Florissant.

ALLANITE. Cerium-Epidote. *Iddings and Cross, Am. Jr. Sc.,* Aug., 1885.
Monoclinic; isomorphous with epidote; crystals either short, flat tabular,
or long and slender, sometimes acicular; twins like those of epidote; also
massive and in angular and rounded grains. Lustre submetallic, pitchy
or resinous—occasionally vitreous. Color pitch-brown to black, either
brownish, greenish, grayish or yellowish. Streak gray, sometimes slightly
greenish or brownish. Subtranslucent to opaque. Fracture uneven or
subconchoidal. Double refraction either distinct or wanting. H. 5·5–6.
G. 3·9–4·2. Comp., silica 33, alumina 15, protoxide of iron 15, cerium 21,
lime 11, lanthanium, didymium, yttria, manganese, etc.
Forms an accessory constituent of the biotite porphyrite of
Ten-Mile District, Summit County, in which the crystals occur of a brilliant black color and oily lustre. In the porphyrite of Mount Silverheels, Park County, of a chestnut-brown
color. In the gneiss of the Medicine Bow Range, the quartz
porphyry of the Mosquito Range, and on Eagle River.

ALLOPHANE. Hydrous Silicate of Alumina. *Konig, Proc. Acad. Nat.*
Sci. Phila., 1876.
Amorphous; usually in thin incrustations, with mammillary surface;
sometimes stalactitic; occasionally almost pulverulent. Lustre vitreous
to subresinous; bright and waxy internally. Color pale sky-blue, sometimes greenish to deep green, brown, yellow or colorless. Streak uncolored. Translucent. Fracture imperfectly conchoidal and shining to
earthy. Very brittle. Adheres to the tongue. H. 3. G. 1·85. Comp.,
silica 22, alumina 33, water 41. An analysis by Genth of what was supposed to be chrysocolla, from Bergen Park, gave 33.85 alumina, and 5·40
of copper oxide, which corresponds to allophane and chrysocolla in the
ratio of 5:1.
Forms a thin bluish crust on limonite near Bergen rauch,
Jefferson County.

ARFVEDSONITE. Soda Hornblende. *Konig, Am. Phi. Soc.*, 1877.
Probably monoclinic; occurs in crystals and cleavable masses. Color
pure black; in thin scales deep green to brown. Streak grayish-green.
H. 6. G. 3·44. Comp., silica 50·5, sesquioxyd of iron 26·9, protoxyd of
iron 12·1, soda 10·5. Analysis of El Paso County mineral by Konig was
published in the proceedings of the American Philosophical Society of
Philadelphia.

Occurs at St. Peter's Dome, El Paso County, with astrophyllite, zircon, etc.

GAHNITE. Zinc-Spinel. *Genth, Am. Phi. Soc.*
Isometric; occurs in octahedrons and dodecahedrons. Lustre vitreous or
somewhat greasy. Color dark green.grayish-green, deep leek-green, greenish-black, bluish, black, yellowish, grayish-brown. Subtranslucent to
opaque. H. 7·5–8. G. 4–4·6. Comp., alumina 61·3, oxyd of zinc 38·7.

Large rough crystals occur in the Cotopaxi mine, Chaffee
County.

TOPAZ. *Cross and Hillebrand, Am. Jr. Sc., Oct.*, 1882; *Bul.* 20, *U. S. Geo.
Sur. W. B. Smith, Bul. U. S. Geo. Sur.*
Orthorhombic; primary form right rhombic prism; crystals usually hemihedral, the extremes being unlike. Cleavage perfect at right angles to the
principal axis; also occurs fine columnar, granular, coarse or fine. Lustre
vitreous. Color straw-yellow, wine-yellow, white, grayish, greenish, bluish, reddish. Transparent to subtranslucent. Pyroelectric. H. 8. G.
3·4–3·65. Analysis of Florissant specimen by Hillebrand: silica 33·15,
alumina 57·01, fluorine 16·04, oxygen for fluorine 6·75.

Found in Crystal Park, south of Manitou, in cavities in
granite containing feldspars, smoky quartz, zircon and phenacite; at the main amazonstone locality near Florissant: and,
more plentifully and in better form than elsewhere, on Devil's
Head Mountain, Douglas County. In small crystals in the
nevadite from Chalk Mountain, where Lake, Eagle and Summit Counties join. A fragment of a crystal 3½ in. in its longest diameter, was found near Florissant, which evidently
came from a crystal about one foot in diameter. It is clear
in parts and has a decided greenish tinge. It was supposed
to be fluor spar by the original collectors, and the other pieces
of the crystal are undoubtedly lost. Crystals are also found
in this locality deposited upon or partially imbedded in amazonstone, albite, etc. In size the crystals vary from a length
of nearly two inches to those which are almost microscopic.
Some of the crystals have a decided greenish tinge, although
many are colorless. The topaz found at Devil's Head Mountain also occur in cavities with amazonstone, smoky quartz,
and other minerals, and are the most noteworthy crystallized
species, some of the specimens found being probably the best
yet discovered in the United States. Much of the topaz is
reddish, though wine-yellow, milk-blue, and colorless crystals
are found. Quite perfect and clear crystals have been found
weighing from five to six ounces.

KALINITE. Potash Alum. Native Alum.
Isometric; usually fibrous or massive, or in mealy or solid crusts. Lustre vitreous. Color white. Transparent—translucent. H. 2-2·5. G. 1·75. Comp., sulphate of potash 18·4, sulphate of alumina 36·2, water 45·5.

Occurs as an efflorescence on argellaceous rocks at Canon City, on Turkey Creek, Little Thompson Creek, and in small quantities in various localities along the "hog-back."

STIBNITE. Antimony Glance. Sulphuret of Antimony.
Orthorhombic; primary form a right rhombic prism ; occurs crystallized in variously modified and terminated rhombic prisms, which are sometimes closely aggregated laterally; lateral planes of crystals deeply striated longitudinally ; also occurs columnar, coarse or fine; also granular to impalpable. Color and streak lead-gray ; inclining to steel-gray; subject to blackish tarnish. Lustre metallic. Sectile. Thin laminæ a little flexible. H. 2. G. 4·51–4·62. Comp., sulphur 28·2, antimony 71·8.

Occurs crystallized in the North Star mine, Sultan Mountain, near Silverton.

QUARTZ.
Rhombohedral; occurs in hexagonal prisms, sometimes terminated at both ends by six-sided pyramids. Colorless when pure; various shades of yellow, red, brown, green, blue, violet and black. Transparent to opaque. Lustre vitreous, sometimes inclining to resinous. Fracture conchoidal. Tough, brittle, friable. H. 7. G. 2·5–2·8. Comp., pure silica, or, oxygen 53·33, silicon 46·67.

Quartz is a constituent of many rocks, and composes most of the pebbles of gravel beds. It takes on more forms and colors than any other mineral, and is divided into three varieties, the vitreous, chalcedonic and jaspery. Vitreous varieties are distinguished by their glassy fracture ; chalcedonic varieties by having a subvitreous or waxy lustre, and generally translucent ; the jaspery varieties are opaque.

ROCK CRYSTAL.
Pure pellucid quartz. Usually occurs in six-sided crystals.

Clear crystals, some of which are doubly terminated, are found near Maysville. Large crystals in the Elk mountains. Pellucid water-worn nodules in the gravel of Platte River. Common as drusy incrustations in the mines. Rock crystal is made into lenses for spectacles, and is known as Scotch and Brazilian Pebbles. The so-called Alaska and California diamonds are cut from rock crystal.

AMETHYST.
Quartz of a clear purple or violet-blue color, of various degrees of intensity. The color is supposed to be derived from manganese.

Fine specimens have been met with in the veins near Dumont. Small crystals at Nevada and Black Hawk. On Grape Creek, Fremont County. In the Elk mountains. In geodes on the Rio Grande River. Occasionally in South Park, associated with smoky quartz, amazonstone and gothite.

ROSE QUARTZ.
A transparent, or nearly transparent variety of quartz, of a rose-red or pink color. It usually occurs massive, and often much fractured.

Occuas in large masses at the summit of Floyd Hill; near Central City; on Bear and Soda Creeks.

YELLOW QUARTZ. False Topaz. Citrine.
Color lemon-yellow, golden, or wine-yellow.

Found in the gravel on the divide between the Platte and Arkansas, about the head of Plum Creek. On the west slope of Pike's Peak, near its base.

CAIRNGORM. Smoky Quartz. Morion.
Color various shades of brown, passing into black.

Abundant about Florissant, El Paso County, where it occurs in "pockets" in the granite, associated with amazon-stone, albite, gothite and fluorite. Some of the crystals are of great size and extreme beauty. One crystal was found which measured 4½ feet in length and 10 inches in diameter at the base. A great many have been found from 20 to 30 inches in length. The crystals are usually transparent, and have been used quite extensively in the manufacture of jewelry. Smoky quartz is also found at the head of Plum Creek; in the neighborhood of Larkspur, Summit and Monument; at the head of Beaver Creek; on Cheyenne mountain; Elk Creek, and in the nevadite of Chalk mountain.

JASPER.
Impure opaque colored quartz. The most common colors are brown, red and yellow. Called *Ribbon Jasper* when the colors are in broad stripes or bands. *Egyptian Jasper* when the colors appear in zones.

Red and yellow jasper is found very plentifully in Middle and South Parks. A large proportion of the petrified wood on the Platte and Arkansas divide, is jasper. Along the sources of Cherry, Kiowa, and Bijou Creeks, it is quite common to find large trees, 60 or 70 feet in length, changed to jasper. *Ribbon Jasper* is found near Larkspur; also in Middle Park, at the junction of the Willow and Grand, in dark green and yellow stripes. *Egyptian Jasper* near Larkspur.

CERARGYRITE. Horn Silver. Chloride of Silver.
Isometric; occurs crystallized in small cubes and acicular prisms; generally massive and looking like wax; sometimes columnar, or bent columnar; often in crusts. Lustre resinous, passing into adamantine. Color pearl-gray, grayish-green, whitish, rarely violet-blue, colorless sometimes when perfectly pure; brown or violet-brown on exposure. Streak shining. Transparent—feebly translucent. Fracture somewhat conchoidal. Sectile. Comp., Chlorine 24.7, silver 75.3.

Common in the mines of Leadville, Silver Cliff and Rosita. Occasionally found near the surface of veins about Georgetown and elsewhere.

CELESTITE. Fibrous Heavy Spar. Sulphate of Strontia.
Orthorhombic; in modified rhombic prisms; crystals sometimes flattened; often long and slender; also fibrous and radiated; sometimes globular; occasionally granular. Lustre vitreous, sometimes inclining to pearly. Color white, often bluish, and sometimes reddish. Streak white. Transparent—subtranslucent. Brittle. Trichroism sometimes very distinct. Phosphoresces when heated. H. 3–3·5. G. 3·92–3·95· Comp., sulphuric acid 43·6, strontia 56·4.

Occurs as nodules in the Garden of the Gods, which, upon being broken, present a mass of beautiful blue crystals; occasionally in fibrous masses. Celestite is used in the arts for making nitrate of strontia, which is employed for producing a red color in fire works.

SPHALERITE. Blende. Zinc-Blende. Sulphuret of Zinc.
Isometric; tetrahedral; occurs crystallized and amorphous; botryoidal and other imitative shapes; sometimes fibrous and radiated; also massive, compact. Cleavage, dodecahedral, highly perfect. Color brown, yellow, black, red, green; white or yellow when pure. Lustre resinous to adamantine. Streak white to reddish-brown. Transparent to translucent. Fracture conchoidal. Brittle. H. 3·5–4. G. 3·9–4·2. Comp., sulphur 33, zinc 67.

Common in many of the gold and silver mines, and frequently rich in gold or silver, although usually of but small value. Fine cabinet specimens of black blende in the Maine and other mines about Georgetown; in the Coaley, Calhoun and Delaware lodes, Gilpin County; beautiful green crystals in the Little Giant lode, near Lawson, Clear Creek County.

MOLYBDENITE. Sulphuret of Molybdena.
Hexagonal; occurs in flat hexagonal tables, with a cleavage parallel to their terminal planes; generally massive with a foliated structure, or in scales. Color pure lead-gray. Lustre metallic. Streak similar to color, slightly inclining to green. Laminæ flexible, but not elastic. Sectile, and almost malleable. Leaves a gray trace on paper, a greenish trace on porcelain. H. 1–1·5, being easily impressed by the nail. G. 4·4–4·8. Comp., sulphur 41, molybdenum 59.

About Georgetown; Rock Creek, Gunnison County; Boulder, Gilpin and Summit counties. Salts of molybdenum are used to some extent in chemical operations.

HALITE. Common Salt. Rock Salt. Muriate of Soda. Chlorid of Sodium.
Isometric; usually occurs in cubical crystals; rarely in octahedrons; faces of crystals sometimes cavernous. Massive and granular, rarely columnar. Cubic cleavage. Lustre vitreous. Color white, sometimes yellowish, reddish, bluish, purplish; often colorless. Transparent—translucent. Fracture conchoidal. Rather brittle. Soluble. Taste purely saline. H. 2·5. G. 2·1–2·257. Comp., chlorine 60·7, sodium 39·3, commonly mixed with some sulphate of lime, chlorid of calcium, etc.

Occurs in solution and as an incrustation about the salt springs in South Park. Works were erected about 1868, and considerable salt manufactured, but the company got into litigation, and the works were abandoned.

GYPSUM. Sulphate of Lime.

Monoclinic; crystals usually in right rhomboidal prisms, with beveled sides; frequently the shape of an arrow-head; eminently foliated in one direction, and cleaving easily, affording laminæ that are flexible but not elastic; crystallized variety called *selenite.* Also occurs in· laminated masses; fibrous, with a satin lustre (*satin-spar*); in stellated or radiating forms, consisting of narrow laminæ; also granular and compact (*alabaster*). Color usually white; sometimes gray, yellow, reddish, brown, black. Streak white. Transparent—opaque. H. 1·5–2. G. 2·3. Comp., sulphuric acid 46·5, lime 32·6, water 20·8.

Occurs in large beds of great purity in the South Park and along the base of the mountains; specially available at Morrison and Colorado Springs, where mills have been built for the manufacture of gypsum into plaster-of-paris.

SELENITE. Crystallized variety of Gypsum.

Occurs either in distinct crystals, or broad folia, the folia sometimes a yard across, and transparent throughout. Cleaves easily, the plates bending in one direction.

Occurs in grouped and single crystals on Box Elder Creek, Laramie County; at River Bend, on the K. P. Railway, in fine crystals; on Bear Creek, near Morrison. Selenite, in thin folia, was formerly extensively used in windows, and is still used for that purpose by the Mexicans in the Socorro mountains.

SATIN SPAR. Fibrous Gypsum.

Occurs in fine fibrous masses, having a satin lustre, and the pearly opalescence of the moonstone.

Fine white masses at Morrison. Pink and white near Quick's ranch, on Plum Creek. Satin Spar is used in the manufacture of beads and other ornamental articles.

ALABASTER. Granular Gypsum.

Fine-grained, and either white or delicately shaded.

On Box Elder Creek, in immense deposits. At Morrison, white and mottled. At Colorado Springs, pink and white. At Canon City, pure white.

GADOLINITE. *Eakins, Proc. Colo. Sc. Soc.,* 1885.

Orthorhombic; in oblique rhombic prism; occurs also in small masses. Color black, greenish-black; dull externally, internally shining; in thin splinters nearly transparent, and grass-green to olive-green. Streak greenish-gray. Lustre vitreous. Subtranslucent—opaque. Fracture conchoidal. H. 6·5–7. G. 4–4·5. Analysis of Colorado mineral by Eakins: Silica 22·13, alumina 2·34, protoxyd of iron 1·13, thoria 0·89, oxide of cerium 11·10, oxide of lanthanum and didymium 21·23, erbium 12·74, yttria 9·50, oxide of iron 10·43, beryllium 7·19, lime 0·34, magnesia 0·14, potassa 0·18, soda 0·28, water 0·86.

Occurs in small fragments on Devil's Head mountain, Douglas County. This is the first observed occurrence of the mineral in this country, and was found by Mr. W. B. Smith, of the U. S. Geological Survey.

CALAMINE. Silicate of Zinc.
Orthorhombic; occurs in obtuse rhombohedrons, and in long quadrilateral tables; also stalactitic, mammilated, botryoidal, fibrous, massive and granular. Lustre subpearly, sometimes adamantine. Color white; sometimes with a delicate bluish or greenish shade; also yellowish to brown. Streak white. Transparent—translucent. Brittle. Fracture uneven. Pyroelectric. Double refraction strong. H. 4·5–5. G. 3·16–3·49. Comp., silica 25·0, oxyd of zinc 67·5, water 7·5.

Occurs as a delicate white incrustation in the mines of Leadville.

KOBELLITE. *H. F.* and *H. A. Kellar, Am. Chem. Jr., vol.* 2.
Radiated structure. Color dark lead-gray. Lustre bright. Soft. G. 6·29. Analysis of mineral from Lillian Co's mines: sulphur 15·21, bismuth 32·62, lead 43·94, silver 5·78, copper trace.

Occurs in the mines of the Lillian company on Printer Boy Hill, near Leadville, in nodules of various sizes up to several feet in diameter, more or less oxidized. The fresh mineral has a fine-grained crystalline structure, and steel-gray color.

GRAPHITE. Plumbago. Black Lead.
Hexagonal; in flat six-sided tables. Commonly in imbedded, foliated, or granular masses. Perfect cleavage. Color iron-black to dark steel-gray. Streak black and shining. Lustre metallic. Soils paper. Sectile; thin laminæ flexible. Feel greasy. H. 1–2. G. 2·0891. Comp., pure carbon, with often a little oxyd of iron mechanically mixed.

At Pitkin, in beds 2 feet thick; large masses at Trinidad; small veins at Maysville, but very impure in each locality. Graphite is largely employed under the name of Plumbago, or Black Lead, for the manufacture of crucibles and other refractory articles; foundry facings, lubricating compounds, electrical supplies, stove-polish, pencil leads and pigments. It has an average spot value of 8 cents a pound.

SCHEELITE. Tungstate of Lime. *Silliman, Am. Jr. Sc.,* June, 1877.
Tetragonal; primary form a right square prism; occurs in attached and imbedded four-sided pyramids, approaching nearly to the octahedron; also occurs reniform with columnar structure, and massive granular. Lustre vitreous inclining to adamantine. Color white, yellowish-white to orange. yellow, brownish, greenish. Streak white. Transparent to translucent, H. 4·5–5. G. 5·9. Comp., lime 19·4, tungstic acid 80·6.

Gold-bearing scheelite occurs in the Golden Queen mine, Lake County.

ROSCOELITE. Silicate of Vanadium. *Genth, Am. Phi. Soc., Phila.,* 1877.
Micaceous in structure; scales minute, often arranged in stellate or fan-shaped groups. Basal cleavage perfect. Color dark clover-brown to greenish-brown. Analysis of Boulder County mineral by Genth: silica 56·74, alumina 19·62, vanadium 7·78, iron oxide 3·84, magnesia 2·63, potassa 8·11, with traces of manganese, soda, lithium and water.

In the mines of Magnolia District, Boulder County, as a thin earthy incrustation of a grayish to olive-green color on calaverite; also inclosed in quartz, and giving it a green color.

HESSITE. Telluric Silver. *Silliman, Am. Jr. Sc., July,* 1874. *Genth, Am. Phi. Soc.,* 1874.

Orthorhombic; massive; compact or fine-grained; rarely coarse granular. Lustre metallic. Color between lead-gray and steel-gray. Sectile and malleable, laminating into thin scales under the pestle, leaving on the agate surfaces metallic streaks of plumbago-like color. H. 2–3·5. G. 8·3 8·6. Analysis by Genth of mineral from the Red Cloud mine, Boulder County: gold 0·22, silver 59·91, copper 0·17, lead 0.45, zinc trace, iron 1·35, tellurium 37·86. An auriferous variety gave: gold 13·09, silver 50·36, tellurium 34·91.

Occurs abundantly in the Slide and Prussian mines, and sparingly in other mines at Gold Hill and Sunshine, Boulder County.

WOLFRAMITE. Tungstate of Manganese.

Orthorhombic; primary form a right rhombic prism; occurs massive and crystalized; irregular lamellar; coarse divergent columnar; massive granular, the particles strongly coherent. Lustre submetallic. Color dark grayish or brownish black. Streak dark reddish brown to black. Sometimes weak magnetic. H. 5–5·5. G. 7·1–7·55. Comp., tungstic acid 75·92, protoxide of iron 19·35, protoxide of manganese 4·73.

Hillebrand mentions its occurrence in the Missouri mine, Hall Valley, Park County. Occurs near Boulder.

URANINITE. Pitchblende. Protoxyd of Uranium. *Am. Jr. Sc.,* 1873.

Isometric; usually massive and botryoidal; also in grains. Structure sometimes columnar, or curved lamellar. Lustre submetallic to greasy or pitch-like, and dull. Color grayish, greenish, brownish, velvet-black. Streak brownish, a little shining. Fracture conchoidal, uneven. H. 5·5. G. 6·4–8. Comp., protoxyd of uranium 32·1, sesquioxyd 67·9, but analyses vary much through mixture with other substances.

A pocket of uraninite was encountered in the Wood lode, Central City, in 1873, nearly all of which was shipped to England by Richard Pearce, esq., of the Boston and Colorado Smelter, and sold for $1.50 per pound. Thin seams of the mineral occur in Jefferson county. A mass of uraninite was encountered in the Jo Reynolds mines, Clear Creek County, in 1885, but unfortunately, it was mixed with the silver ore from the mine, and milled, the adulteration causing a serious loss to the miners. But two or three specimens were saved, one of which is in the collection of the compiler, also specimens from the Wood lode.

Uranium has the color and lustre of silver, but is harder, and gives out sparks when struck with a hammer. It oxidizes gradually when exposed to the air, burns when heated on platinum-foil, and is dissolved by nitric acid. Its specific gravity is 18·7. Its compounds are used in considerable quantities in chemical operations and in porcelain painting, affording a yellow or black color, according to the process of baking.

CALAVERITE. Telluride of Gold. *Genth, Am. Phi. Soc., Phila.*, 1877.
Occurs massive and in small crystals. Color bronze-yellow. Streak yellowish-gray. Brittle. Fracture uneven, inclining to subconchoidal. Analysis of Colorado mineral by Genth: tellurium 57·32, gold 38·75, silver 3·03, oxyd of vanadium 0·05, protoxyd of iron 0·03, alumina, magnesia,etc., 0·55, quartz 4·96. Theoretical formula, tellurium 57·93, gold 39·01, silver 3·06.

In the Slide, Keystone, Mountain Lion and other mines, in Boulder County. Small imperfect crystals imbedded in quartz.

FERROTELLURITE. (*new.*) *Genth, Am. Phi. Soc. Phil.*, 1877.
In delicate radiating tufts; also in very minute prismatic crystals. Color between straw and lemon-yellow, inclining to greenish-yellow. Contains iron and tellurium.

Found as a coating on quartz associated with native tellurium and tellurite, at the Keystone mine, Magnolia District, Boulder County.

LIONITE. (*new.*) Variety of Native Tellurium. *Genth, Am. Phi. Soc. Phil.*, 1877.
Occurs in thin plates. Color dark gray. H. 3. G. 4·005. Analysis of Colorado mineral by Genth : tellurium 55·54, quartz 35·91, alumina and iron oxyd 6·14, manganese and iron oxide 0·19, calcium and iron oxide 0·26, gold 1·53, silver 0·25.

Occurs in the Mountain Lion mine, Boulder County, from which it was named by Berdell.

MAGNOLITE. (*new*). Telluride of Mercury. *Genth, Am. Phi. Soc.*, 1877.
Occurs in radiating tufts of very minute acicular or capillary crystals. Color white. Lustre silky. Contains mercury and tellurium.

Occurs as a product of the decomposition of coloradoite in the upper part of the Keystone mine, Magnolia District, Boulder County. Named from Magnolia District.

TELLURITE. Tellurious Acid. *Genth, Am. Phi. Soc. Phil.*, 1877.
Occurs in prismatic crystals and in spherical masses, radiated in structure; also as an incrustation. Crystals cleavable in one direction. Color yellow to white.

In the John Jay mine, Boulder County, as an incrustation in cracks in native tellurium, and in minute prismatic crystals.

CHRYSOCOLLA. Silicate of Copper.
Cryptocrystalline; often opal-like or enamel-like in texture; earthy; incrusting or filling seams; sometimes botryoidal. Color various shades of blue, passing into green. Streak, when pure, white. Translucent to opaque. Fracture conchoidal. H. 2–4. G. 2·2. Comp., silica 34·2, oxyd of copper 45·3, water 20·5. An analysis by Genth of what was supposed to be chrysocolla, from Bergen Park, gave 33.85 alumina, and 5·40 of copper oxide, which corresponds to allophane with probably some chrysocolla.

Found in copper lodes on Bear Creek; near Canon City ; at the head of San Luis Valley ; in the Champion lode, Idaho Springs, and in small quantity in many other localities.

ERYTHRITE. Cobalt Bloom. Red Cobalt Ore. Cobalt Ochre. *W. F. Hillebrand, Am. Jr. Sc.*, May, 1884; *Proc. Col. Sic. Soc.*, Vol. I.

Monoclinic; in acicular diverging crystals, modified at the edges, and whose form is a right oblique-angled prism; also found in globular and reniform shapes, having a drusy surface and a columnar structure; sometimes stellate; also pulverulent and earthy, incrusting. Lustre vitreous to earthy. Color crimson and peach-red, sometimes pearly or greenish-gray; red tints incline to blue perpendicular to cleavage face. Streak a little paler than the color; the dry powder deep lavender-blue. Thin laminæ flexible in one direction. Sectile. H. 1·5–2·5, G. 2·94. Comp., arsenic acid 38·43, oxyd of cobalt 37·55, water 24·02.

A product of the alteration of lollingite in the mines on Teocalli Mountain, Brush Creek, Gunnison County.

AMPHIBOLE. Hornblende.

Monoclinic; crystals often long and bladed, sometimes stout; fibrous, columnar, coarse or fine, fibers often like flax, when it is known by other names. Color between black and white, through various shades of green, inclining to blackish green. This species has numerous varieties, differing much in external appearance and in composition. Silica, alumina, lime, iron and magnesia enter into the composition.

Hornblende is an essential constituent of many of the rocks of the mountainous region, such as hornblende gneiss, a tough dark rock, which is quite abundant, seyenite, trap., etc. It is occasionally found massive and crystallized.

ASBESTOS. A variety of Amphibole.

Occurs in fine fibrous masses, of light color.

In the Star of the West and other lodes at the head of of North Boulder Creek, Boulder County, associated with galena. Near Maysville, Chaffee County, in mineral veins.

Asbestos was woven into cloth by the ancients, which, from its incombustibility, was used to wrap the bodies of the dead before placing them on the funeral pile, by which the ashes were preserved for subsequent preservation in vases. Clothing was manufactured from it, which was cleansed by burning instead of washing. Its chief use now is for making fire-proof roofing, boiler lining and fire proof safes.

TREMOLITE.

The name given to the white or light greenish variety of Amphibole. It occurs in long slender blades, or in columnar and radiated aggregations.

In white, radiated masses near Maysville, Chaffee County.

ACTINOLITE. A variety of Amphibole.

Fibrous, columnary or massive. Color bright green or grayish-green.

On Bear Creek, Jefferson County, in fibrous aggregations, of a dark green color. In the mineral veins at the head of North Boulder Creek, and in the Partridge lode, on Coal Creek, Boulder County. Light green and bluish-green on Mount Ouray. Specimens of the amphibole minerals in the compiler's collection.

COLORADOITE. (*new.*) Telluride of Mercury. *Genth, Am. Phi. Soc., Phila.,* 1877.

Massive ; granular, sometimes imperfectly columnar (due to admixed sylvanite?) Luster metallic. Color iron-black inclining to gray. Fracture uneven to subconchoidal. Theoretical composition, tellurium 39·02, mercury 60·98. Analyses of mineral from the Keystone and Smuggler mines by Genth, varied a great deal on account of an admixture of gold and quartz. That from the Keystone mine carried from 8 to 46 per cent of gold and quartz, with small amounts of alumina, iron, vanadium and magnesia. Mineral from the Smuggler mine contained quartz and gold, 3·05, tellurium 34·49, mercury 48·74, gold 7·67, silver 7·18, iron 0·92, copper 0·16, zinc 0·50. II. 3. G. 8·627.

Occurs sparingly in the Keystone, Smuggler and Mountain Lion mines, Boulder county.

MENDOZITE. Soda Alum.

Occurs in white fibrous and pulverulent masses. Resembles fibrous gypsum, but harder. H. 3. G. 1·88. Reported analysis of Colorado mineral: sulphuric acid 38·35, sulphate of alumina 14·66, sulphate of soda 4·35 water 42·64.

F. F. Chisolmn reports its occurrence in a lode one-and-a-half miles shuth-east of Red Cliff.

WILLEMITE. Anhydrous Silicate of Zinc.

Rhombohedral ; occurs in regular six-sided prisms ; also massive, fibrous and in grains. Color whitish, yellowish, flesh-red, apple-green and dark-brown. Streak same as color. Luster vitreo-resinous. Transparent to opaque. Brittle. Fracture conchoidal. Double refracting. H. 5·5. G. 3·84—4·18. Comp., silica 27·1, oxyd of zinc 72·9.

Occurs sparingly at the head of the Rio La Plata as a pale green crystalline mineral filling cavities in other zinc ores.

ILESITE. (*new.*) Sulphate of Manganese. *M. W. Iles, American Chemical Journal,* 1881.

Occurs in thick prisms which are frequently terminated by truncated pyramids. Friable. Taste bitter, astringent. Color white. Soft. G. 2·16. Analysis by Iles : protoxyd of iron 4·18, oxyd of zinc 5·97, protoxyd of manganese 22·31, sulphuric acid 36·07, water 31·60.

Occurs in veins from two to eight inches wide in the McDonnell lode, Middle Swan Creek, Hall Valley, Park county. Named after Dr. M. W. Iles.

CHALCANTHITE. Blue Vitriol. Sulphate of Copper.

Triclinic ; rarely found in distinct crystals ; generally occurs stalactitic, reniform, amorphous and pulverulent. Color Berlin-blue to sky-blue of various shades ; sometimes greenish ; turns white on exposure. Translucent. Luster vitreous. Taste metallic and nauseous. Comp., sulphuric acid 32·1, oxyd of copper 31·8, water 36·1.

Occurs in a number of mines in Hall Valley, Park county ; in the Whale tunnel, near Idaho Springs ; in a deposit below Black Hawk. It is formed by the decomposition of iron and copper pyrites. When purified it is employed in cotton and linen printing, and for various other purposes in the arts. It is sometimes employed to prevent dry rot in wood.

SCHIRMERITE. (new.) Bismuth Silver. *F. A. Genth, Proc. Am. Phi. Soc., Phila., vol.* 14, 1874.
Massive, without crystalline structure; fine granular. Fracture uneven. Brittle. Color lead-gray, inclining to iron-black. Luster metallic. Soft. G. 6·737. Analysis by Genth: lead 12·76, silver 24·75, bismuth 47·27, zinc 0·13, iron 0·07, sulphur 15·02.

First noticed in the Treasure Vault mine, Geneva district, Clear Creek county, from whence the mineral came which was analyzed by Mr. Genth. It occurs in all the older veins in the district, finely disseminated through quartz. It has never been noticed in the newer or subsequently-formed veins which carry a large amount of galena, very little of which is found in the bismuth-bearing veins. Dana's Mineralogy credits the mineral to the Red Cloud mine, Boulder county, where he says it "occurs with other tellurium minerals." The location is erroneous, as the mineral has never been found in that county. It was named after J. F. L. Schirmer. The schirmerite of Endlich was found to be a mixture of petzite and some other minerals, and therefore not entitled to a name. Specimens in compiler's cabinet.

SYLVANITE. Graphic Tellurium. Telluride of Gold and Silver. *Silliman, Am. Jr. Sc., July,* 1874, *and Clarke, Oct.,* 1877 ; *Genth, Am. Phi. Soc., Phila.,* 1874.
Monoclinic; occurs in indistinct and minute circular crystals, modified at the edges and angles, and often grouped in rows, forming triangular figures like letters; also massive, imperfectly columnar, granular. Cleavage in two directions, nearly at right angles; one very perfect; Luster metallic. Color steel-gray to tin-white and brass-yellow. Streak like color. H. 1·5–2. G. 5·732-8·28. Analysis by Genth of mineral from Red Cloud mine, Gold Hill: gold 24·83, silver 13·05, tellurium 56·31, copper 0·23, zinc 0·45, iron 3·28, sulphur 1·82, selenium trace, quartz 0·32. Analysis of mineral from Grand view mine, by Clarke: tellurium 52·96, gold 26·39, silver 10·55, iron 4·45, sulphur 5·62.

Quite abundant in crystals and crystalline masses disseminated ·through quartz, in the the Red Cloud, Prussian, Cold Spring and other veins at Gold Hill; in the American and Grand View veins at Sunshine, Boulder county. Specimens in compiler's cabinet.

ASTROPHYLLITE. *Konig, Am. Phi. Soc., Phila.,* 1877.
Orthorhombic; usually in tabular prisms; often lengthened into strips with parallel sides; easily cleavable. Luster submetallic, pearly. Color bronze-yellow, gold-yellow, brownish. Powder like mosaic gold. Translucent and slightly elastic in thin leaves. H. 3. G. 3·324. Analysis by Konig of Colorado mineral ; silica 34·68, titanic acid 13·58, zirconia 2·20, sesquioxyd of iron 5·56, alumina 0·70, protoxyd of iron 26·10, protoxyd of manganese 3·48, soda 2·54, potassa 5·01, water 3·54, magnesia 0·30, oxyd of copper 0·42, tantalic acid (?) 0·80.

Beautiful specimens are found in veins on Cheyenne Mountain, in quartz, associated with arfvedsonite and zircon. Specimens in compiler's collection.

GARNET.

Isometric; occurs in dodecahedrons, also in trapezohedrons, and both forms are sometimes variously modified. Cleavage parallel to the faces of the dodecahedron, rather distinct. Also found massive granular and coarse lamellar. Color red, brown, yellow, white, apple-green, black; some red and green colors often bright. Streak white. Lustre vitreous—resinous. Transparent—subtranslucent. Fracture subconchoidal, uneven. H. 6.5-7.5. G. 4·13. Composition varies, silica, alumina, iron, lime, etc. There are three prominent varieties, based on the nature of the composition. (See *Spessartite* and *Almandite*.)

Red, brown and black garnets occur quite frequently in the Archean. A reddish-brown garnet between Georgetown and Green Lake, and on Saxon Mountain. Quite abundant near the head of Fall River, on Trail Creek, Bergen Park, Central City, and other localities. A black variety in a vein near Hoosier Pass. Small garnets, probably *almandite*, of beautiful color, are found in quantities in the gold washings at Fairplay and Breckenridge; impure *almandite* near Salida (*p.48*). *Spessartite* at Nathrop in rhyolite, and in the Grand Canon of the Arkansas (*p. 36*).

LIMONITE. Brown Iron Ore. Brown Ochre.

Usually in stalactitic and botryoidal or mammilary forms, having a fibrous or subfibrous structure; also concretionary, massive, and occasionally earthy. Lustre silky, often submetallic; sometimes dull and earthy. Color of surface of fracture various shades of brown, commonly dark, and none bright; sometimes with a nearly black varnish-like exterior; when earthy, brownish-yellow, ochre-yellow. Streak yellowish-brown. H. 5—5·5. G. 3·6-4. Comp., sesquioxyd of iron 85·6, water 14·4.

A brown hematite of excellent quality is found in the Hot Springs mine, Saguache County, which has been quite extensively mined. At Villa Grove, in the same county. A bog ore is found near Crested Butte. Analyses of ore from these localities have been made at the State School of mines

	Hot Springs.	Villa Grove.	Crested Butte.
Silica	9·33	10·19	2·50
Water	10·51	13·57	23·97
Alumina	3·43	3·60	0·28
Oxide of Manganese	0·35		
Lime	0·83	0·45	0·22
Magnesia...................	0·06	0·22	0·12
Peroxide of Iron	75·23	70·39	72·47
Phosphoric acid...........	0·071	0·92	0·333
Sulphur	0·019	0·18	
	99·830	99·62	99·893

"This ore is a chemical curiosity," says Prof. Chauvenet, of the State School of Mines, in speaking of the hematite from Crested Butte. "containing at once figures on certain constituents which are phenomenal for highness and lowness respectively. After burning this ore would be enriched to nearly sixty-seven per cent of metallic iron by the loss of water and organic matter.'' The ore is said to approximate to half a mile square in extent, with an uncertain depth.

Limonite exists in numerous other localities—between South Boulder and Coal Creeks; on Bear Creek; in South Park; at Trinided; beautiful specimens on Little Thompson, at the foot of the mountains.

NATRON. Carbonate of Soda.
Monoclinic. Vitreous to earthy. White, sometimes gray or yellow. Taste alkaline. H. 1–1·5. G. 1·423. Comp., carbonic acid 26·7, soda 18·8, water 54·5.

About twelve miles south-west of Denver, between Turkey and Bear Creeks, are four small lakes containing in solution a large amount of soda.

MAGNETITE. Magnetic Iron Ore. Octahedral Iron Ore.
Isometric; octahedrons and rhombic dodecahedrons: cleavage octahedral. Massive, and in particles of various sizes, sometimes impalpable. Lustre metallic—submetallic. Color iron-black. Streak black. Fracture subconchoidal, shining. Brittle. Strongly magnetic, sometimes possessing polarity. H· 5·5–6·5. G. 4·9–5·2. Comp., oxygen 27·6, iron 72·4.

The Calumet mine, Chaffee County, is the only iron mine in the State producing magnetite, upon which the steel works at Pueblo are largely dependent. The mine has been opened to a depth of about 300 feet, and to a much greater distance horizontally. Analysis by School of Mines, Golden:
Silica 7·04, alumina 1·90, peroxyd of iron 56·76, protoxyd of iron 26·88, bisulphide of iron 1·14, lime 1·59, magnesia 1·70, phosphoric acid 0·16, titanic acid trace.

At Caribou a seven-foot vein of very pure ore. On Elk Creek.

HEMATITE. Specular Iron. Red Ochre.
Rhombohedral; in complex modifications of a rhombohedron; crystals occasionally thin tabular; also occurs granular, botryoidal, and stalactitic shapes; lamellar, the laminæ variously bent, and thick or thin. Lustre metallic and occasionally splendent; sometimes earthy. Color dark steel-gray or iron-black; in very thin particles blood-red by transmitted light; when earthy, red. Streak cherry-red or reddish-brown. Sometimes attractable by the magnet, and occasionally even magnetic-polar. H. 5·5–6·5. G. 4·5–5·3. Comp., oxygen 30, iron 70.

An impure hematite occurs in the Hawkins bank, Chaffee County. The following analysis of the ore was made at the State School of mines:
Silica 22·33, water 3·10, alumina 3.06, lime 0·08, magnesia 0·06, peroxide of iron 70·40, phosphoric acid 0·614, sulphur 0·056.

Occurs as an ochre near Florissant, with gothite, microcline, quartz, fluorite, etc. A micaceous variety quite abundant on Left Hand Breek, Ward District, Boulder County, and is found in small quantities in numerous localities.

MUSCOVITE. Common Mica. Potash Mica. Oblique Mica.
Orthorhombic; occurs in oblique rhombic prisms with basal cleavage; usually in thinly foliated masses; folia often aggregated in stellate, or globular forms. Lustre more or less pearly. Color white, gray, brown, hair-brown, pale green, violet, yellow, dark olive-green, rarely rose-red; often different for transmitted and reflected light, and different also in vertical and transverse directions. Streak uncolored. Transparent to translucent. Thin laminæ flexible, elastic, tough. Double refracting. H. 2-2.5 G. 2.75-3.1. Comp., silica 46, alumina 37, potash 9, peroxyd of iron 4, fluoric acid 1, water 2.

A constituent of much of the Archean rocks, and occasionally found in quite large masses. Several attempts have been made to develop deposits in Jefferson and Clear Creek Counties, but without success. Near Canon City quite a large deposit which has been worked with limited success.

BIOTITE. Magnesia-Mica. Hexagonal Mica.
Hexagonal; prisms usually tabular, with perfect basal cleavage. Often in disseminated scales, sometimes in massive aggregations or cleavable scales. Lustre splendent, and more or less pearly on a cleavage surface, and sometimes submetallic when black; lateral surface vitreous when smooth and shining. Colors usually green to black, often deep black in thick crystals; thin laminæ green, blood-red, or brown by transmitted light; rarely white. Streak uncolored. Transparent to opaque. Tough and elastic. H. 2.5-3. G. 2.7-3.1. Comp., varies, silica 39, alumina 15, peroxyd of iron 8, magnesia 24, soda 1, potash 9, water 1, flluorine, chlorine, etc.

The mica of the Archean rocks is mainly biotite. It also occurs in the dike rocks.

BISMUTITE. Carbonate of Bismuth.
In implanted acicular crystallizations (pseudomorphs); also incrusting or amorphous; pulverulent. Lustre vitreous when pure; sometimes dull. Color white, mountain-green, and dirty siskin-green; occasionally straw-yellow and yellowish-gray. Streak greenish-gray to colorless. Subtranslucent—opaque. H. 4-4.5. G. 6.86-6.909.

Near Cummens City, North Park. In the mountains west of Fort Collins, where a number of veins have been prospected. In the Las Animas mine, Gold Hill. In white quartz on Guy Hill, Jefferson County, with bismuthinite.

ALABANDITE. Sulphuret of Manganese.
Isometric; in cubes and octahedrons. Cleavage, cubic perfect. crystals sometimes cruciform, made of five combined octahedrons. Usually massive. Lustre submetallic. Color iron-black, tarnished brown on exposure. Streak green. H. 3·5-4. G. 3·95-4·04. Comp., sulphur 36·7, manganese 63·3.

A "pocket" of this mineral was encountered in sinking the shaft on the Queen of the West mine, Summit County, about 300 feet from the surface. Occasionally crystals are met with. The only immediate associate mineral is rhodochrosite, although the mine carries galena, pyrite and argentite. The alabandite carries about 30 ounces of silver to the ton. This is the first published instance of its occurrence in America.

SPESSARTITE. Manganese-Alumina Garnet. *W. Cross, U. S. Geo. Sur., Am. Jr. Sc., June, 1866.*

Isometric; occurs in dodecahedrons, also in trapezohedrons, and both forms are sometimes variously modified. Cleavage parallel to the faces of the dodecahedron, rather distinct. Also found massive granular and coarse lamellar. Color dark hyacinth-red, sometimes with a shade of violet, to brownish-red. Streak white. Lustre vitreous—resinous. Transparent—subtranslucent. Fracture subconchoidal, uneven. H. 6.5–7.5. G. 4·13. Analysis of Nathrop mineral by L. G. Eakins: silica 35.66, alumina 18.55, sesquioxyd of iron .32, protoxyd of iron 14.25, protoxyd of manganese 29.48, lime 1.15, potassa .27, soda .21, water .44.

Occurs at Nathrop in lithophyses of rhyolite, associated with topaz. In a similar rock in the Grand Canon of the Arkansas.

EPIDOTE.

Monoclinic; in right rhomboidal prisms, more or less modified, often with six or more sides; also fibrous, divergent or parallel; also granular, particles of various sizes, sometimes fine granular and forming rock masses. Lustre vitreous, inclining to pearly or resinous. Color pistachio-green to brownish-green, greenish-black to black; sometimes clear red and yellow; also gray and grayish-white. Subtransparent—opaque; generally subtranslucent. Fracture uneven. Brittle. H. 6–7. G. 3.25–3.5. Comp., silica 37.0, alumina 26.6, lime 20.0, protoxyd of manganese 0.6, water 1.8.

Of frequent occurrence in the Archian, usually massive. Very good crystals have been found on Floyd Hill and on Bear Creek.

MALACHITE. Green Carbonate of Copper.

Monoclinic; rarely occurs crystallized; usually massive or incrustating, with surface tuberose, botryoidal or stalactitic, and structure divergent; often delicately compact fibrous, and banded in color; frequently granular or earthy. Lustre of crystals adamantine, inclining to vitreous; fibrous varieties more or less silky; often dull and earthy. Color bright green. Streak pale green. Translucent, subtranslucent, opaque. Fracture subconchoidal, uneven. H. 3.5.—4. G. 3.7—4.01. Comp., carbonic acid 19.9, protoxyd of copper 71.9, water 8.2.

Of general occurrence throughout the State near the surface of mineral veins, but in small quantities, and usually as incrustations.

CERUSSITE. Carbonate of Lead. White Lead Ore.

Orthorhombic; crystals usually thin, broad and brittle; sometimes stout; rarely fibrous; often granular, massive and compact; sometimes stalactitic. Lustre adamantine, inclining to vitreous or resinous; sometimes pearly; frequently submetallic if the colors are dark, or from a superficial change. Color white, gray, grayish-black, yellowish, sometimes tinged blue or green by some of the salts of copper. Streak uncolored. Transparent—subtranslucent. Fracture conchoidal. Very brittle. H. 3–3·5. G. 6.465–6.480. Comp., carbonic acid 16.5, oxyd of lead 83.5.

Abundant at Leadville, and occasionally in fine crystalline masses. Also abundant in numerous localities south of Leadville. Occurs in small quantities at the surface of silver veins in the Archian.

TITANITE. Sphene.

Monoclinic; in very oblique rhombic prisms; crystals usually thin, with sharp edges. Cleavage in one direction sometimes perfect. Occasionally massive. Color grayish-brown, gray, brown or black; sometimes yellow or green; streak uncolored. Lustre adamantine to resinous. Transparent to opaque. Brittle. H. 5–5·5. G. 3·4–3·56. Comp., silica 30·5, titanic acid 41·3, lime 28·2.

Occurs in crystals in the augitic rock on Italian Mountain, Gunnison County.

CHRYSOLITE. Olivene. Peridot.

Orthorhombic; in right rectangular prisms, having perfect cleavage parallel with the smaller lateral plane; usually in imbedded grains, of an olive-green color, looking like green bottle glass; also yellowish-green. Transparent to translucent. Looks much like glass in the fracture, except in the direction of the cleavage. H. 6–7. G. 3·33–3·5. Comp., silica 38·5, magnesia 48·4, protoxyd of iron 11·2, oxyd of manganese 0·3, alumina 0·2.

Occurs in the basalt of Bastion Peak.

PYROMORPHITE. Green Lead Ore. Phosphate of Lead.

Hexagonal; occurs in six-sided prisms, usually modified on the edges, and frequently striated horizontally; often globular, reniform, and botryoidal, with a subcolumnar structure; also fibrous and granular. Lustre resinous. Color green, yellow and brown of different shades; sometimes wax-yellow and fine orange-yellow; also greenish-white or milk-white. Streak white, sometimes yellowish. Subtransparent–subtranslucent. Brittle. H. 3·5–4. G. 6·5–7·1. Comp., phosphoric acid 15·7, oxyd of lead 74·1, chlorine 2·6, lead 7·6

Sparingly in the mines at Leadville, and in the Astor and Freeland mines, Clear Creek County.

NEPHELITE. Hexagonal; occurs in six and twelve sided prisms, with plane or modified summits. Prismatic cleavage, distinct. Also massive, compact, and thin columnar. Lustre vitreous, greasy; a little opalescent in some varieties. Colorless, white or yellowish; also when massive, dark green, greenish or bluish-gray, brownish and brick-red. Transparent–opaque. Fracture subconchoidal. H. 5·5–6. G. 2·5–2·65. Comp., silica 44·2, glucina 33·7, soda 16·9, potash 5·2.

In the basalt of Bastion Peak and elsewhere in the Elkhead Mountains.

ANNABERGITE. Nickel Ochre. Nickel Green. Arseniate of Nickel.

Monoclinic; in capillary crystals; also massive and disseminated. Color fine apple-green. Streak greenish-white. Fracture uneven, or earthy. Soft. Comp., arsenic acid 38·6, oxyd of nickel 37·2, water 24·2.

Occurs as an alteration product of niccolite in the Gem mine, on Pine Creek, Fremont County.

CHALCOCITE. Vitreous Copper Ore. Copper Glance.

Orthorhombic; in six-sided prisms, variously modified; also massive, structure granular, or compact and impalpable. Lustre metallic. Color and streak blackish lead-gray; often tarnished blue or green. Fracture conchoidal. Streak sometimes shining. H. 2·5–3. G. 5·5–5·8. Comp. sulphur 20·2, copper 79·8.

Abundant in copper veins near Canon City; also in Bergen Park and on Bear Creek, Jefferson County.

NAGYAGITE. Black Tellurium. Foliated Tellurium.

Tetragonal; occurs in small six-sided tables with a basal cleavage; also massive-granular, but generally foliated. Lustre metallic, splendent. Streak and color blackish lead-gray. Flexible in thin laminæ. H. 1–1·5. G. 6·85–7·2. Comp., tellurium 15 to 32, sulphur 3 to 9, lead 50 to 60, gold 5 to 12.

Occasionally met with in the tellurium mines of Boulder County.

GALENITE. Galena. Sulphuret of Lead.

Isometric; occurs crystallized in the cube, octahedron, and in numerous combinations of these, with planes of other figures; also in amorphous masses with a lamellar structure; frequently granular; occasionally fibrous. Cleavage cubic. Lustre metallic. Color and streak pure lead-gray. Surface of crystals occasionally tarnished. Sectile. Easily pulverized, and easily frangible. H. 2·5–2·75. G. 4·25–7·7: Comp., sulphur 13·4, lead 86·6.

Of general occurrence throughout the mining regions, and usually containing a profitable per cent of silver. Good crystals only occasionally met with. Specially fine clusters have been taken from the Dunderberg and Terrible mines, Georgetown, and from the Running, Calhoun and Glennan mines, in Gilpin County.

DECHENITE.

Occurs massive, botryoidal, nodular, stalactitic; sometimes traces of a columnar structure. Lustre of fresh fracture greasy. Color fine deep red to yellowish-red and brownish-red; also leather-yellow. Streak orange-yellow to pale yellow. H. 3–4. G. 5·6–5·81. Comp., vanadic acid 47, oxyd of lead 53.

Occurs in the Evening Star mine, Leadville, of dark red to yellow color.

WITHERITE. Carbonate of Baryta.

Orthorhombic; in modified rhombic prisms and in six-sided prisms terminated with pyramids; also in globular, tuberose and botryoidal forms; structure columnar, granular or amorphous. Lustre vitreous, inclining to resinous, on surface of fracture. Color white, often yellowish or grayish. Streak white. Subtransparent—translucent. Fracture uneven. Brittle. H. 3–3·75. G. 4·29–4·35. Comp., carbonic acid 22·3, baryta 77·7.

In white masses on Cotton Creek, San Luis Park. This mineral is poisonous.

BARITE. Heavy Spar. Sulphate of Baryta.

Orthorhombic; in modified rhombic and rectangular prisms; crystals usually tabular; also columnar, fibrous, granular and compact. Lustre vitreous, inclining to resinous, sometimes pearly. Streak white. Color white, yellow, gray, blue, red, brown. Transparent, translucent, opaque. Sometimes fetid when rubbed. H. 2·5–3·5. G. 4·3–4·8. Comp., sulphuric acid 34, baryta 66.

Of frequent occurrence in the mines, both crystallized and massive. Extensive beds on Vasquez River. Fine single crystals with calcite on Apishapa Creek. Beautiful clusters of crystals in the Seven-Thirty mine, Georgetown. Barite is used as a paint, and in adulterating white lead.

CALCITE. Calcareous Spar. Carbonate of Lime.

Rhombohedral; occurs crystallized in upwards of 800 varieties of form. The primary form is an obtuse rhombohedron, which may readily be obtained by cleavage. Also fibrous, both coarse and fine; sometimes lamellar; often granular; also stalactitic. Lustre vitreous—subvitreous—earthy. Color white or colorless, also various shades of red, green, gray, blue, yellow, violet; brown and black when impure. Streak white or grayish. Transparent—opaque. Double refraction strong. G 2·508–2·776. H. 2·6–3·5. Comp., carbonic acid 44, lime 56.

Large white masses on Bear Creek, five miles above Morrison, in granite. On Box Elder Creek, Laramie County, in aggregations of crystals, and in many other localities along the base of the mountains.

ICELAND SPAR.

Transparent variety of calcite. Double refracting in a high degree.

Near Canon City, on Beaver Creek; on the Greenhorn and Apishapa; in the lava of the Table Mountains, Golden.

ARGENTINE.

A foliated, white pearly calcite.

On Chalk Mountains, in veins sometimes from two to three feet thick.

MARBLE. Granular Limestone.

Calcite having a texture like loaf sugar when broken. The texture varies from coarse to fine granular, and the latter passes by imperceptible shades into compact limestone. The colors are various and usually clouded.

A large vein of black white and gray marble in South Park. Brownish-gray marble near Canon City. Hard compact limestone frequently makes a fine stone when polished. A black variety veined with white occurs on Cotton Creek, San Luis Park. Breccia marble near Boulder. Shell marble near Colorado Springs. Encrinal marble at Sangre de Christo Pass.

DESCLOIZITE.

Orthorhombic. Lustre bright. Color black to olive-brown; smallest crystals olive-green, with a chatoyant bronze lustre; by transmitted light along the edges light brown, inclining to red; on a surface of fracture, colors zoned with straw-yellow, reddish-brown and black; nearly clear at middle and darkest at extremities of crystal. H. 3·5. G. 5·839. Comp., vanadic acid 29·3, oxyd of lead 70·7.

Sparingly in the Evening Star and other mines, Leadville.

KAOLINITE. Porcelain Clay.

Orthorhombic; occurs in rhombic, rhomboidal, or hexagonal scales or plates; sometimes in fan-shaped aggregations; usually constituting a clay-like mass, either compact, friable or mealy. Lustre of plates pearly; of mass, pearly to dull earthy. Color white, grayish-white, yellowish, sometimes brownish, bluish or reddish. Scales flexible. Usually unctious and plastic. H. 1–2·5. G. 2·4–2·63. Comp., silica 46, alumina 40, water 14.

Occurs as the result of the decomposition of the feldspars of granitic rocks and porphyries throughout the state. Kaolin is used in making porcelain and china ware.

VESUVIANITE. Idocrase.

Tetragonal ; crytals usually in modified square prisms ; also found massive. Color brown to green, the latter frequently bright and clear; occasionally sulphur-yellow, and also pale blue. Lustre vitreous, often inclining to resinous. Streak white. Subtransparent to nearly opaque. H. 6·5. G. 3·34. Comp., silica 38·4, alumina 23·7, protoxyd of iron 4·0, lime 29·7, magnesia and protoxyd of manganese 5·2.

Occurs in beautiful yellowish-green crystals on Italian Mt., Gunnison County, which are said to be the finest found in the United States. In brown crystals on Bear Creek.

AZURITE. Chessy Copper. Blue Carbonate of Copper.

Monoclinic; in modified oblique rhombic prisms, the crystals rather short and stout; lateral cleavage perfect. Also massive, incrusting and earthy. Lustre vitreous, almost adamantine. Color various shades of azure-blue, passing into berlin-blue. Streak blue, lighter than the color. Transparent–subtranslucent. Fracture conchoidrl. Brittle. H. 3·5–52·5. G. 3·5–3·831. Comp., carbonic acid 25·6, oxyd of copper 69·2, water 5·2.

Occurs as on incrustation on the surface ores of many of the mines throughout the state.

ARSENOPYRITE. Mispickel. Arsenical Pyrites.

Orthorhombic; crystallizes in right rhombic prisms, parallel to whose planes it may be cleaved ; also occurs columnar, straight and divergent; granular or compact. Lustre metallic. Color silver-white, inclining to steel-gray. Streak dark grayish-black. Fracture uneven. Brittle. H. 5·5–6. G. 6·0–6·4. Comp., arsenic 46·0, sulphur 19·6, iron 44·4. Part of the iron sometimes replaced by cobalt.

Occurs sparingly in some of the mines of Gilpin and Clear Creek Counties.

CUPRITE. Red Oxyd of Copper.

Isometric; in regular octahedrons and modified forms of the same. Cleavage octahedral. Also massive, granular; sometimes earthy. Lustre adamantine or submetallic to earthy. Color red, of various shades, particularly cochineal-red ; occasionally crimson-red by transmitted light. Streak brownish-red, shining. Subtransparent–subtranslucent. Fracture conchoidal, uneven. Brittle. H. 3·5–4. G. 5·85–6·15. Comp., oxygen 11·2, copper 88·8.

In crystals, Sacramento Gulch, and in Sweet Home mine, Buckskin. Malachite lode, Bear Creek. Massive at Poncha Springs.

OPAL.

Massive, amorphous; sometimes small reniform, stalactitic, or large tuberose. Lustre vitreous, frequently subvitreous; often inclining to resinous, and sometimes to pearly. Color white, yellow, red, brown, green, gray, generally pale ; sometimes a rich play of colors, or different colors by refracted or reflected light. Transparent to nearly opaque. H. 5·5–6·5. G. 1·9–2·3. Comp., silica, same as quartz, with a little water.

A milk-white opal occurs in the decomposition product of the lava on Buffalo Peaks, in the Mosquito Range. A yellowish, slightly translucent opal is quite abundant six miles south of the salt works, in South Park, associated with jasper and chalcedony.

FIRE OPAL.

A variety of opal presenting hyacinth-red to honey-yellow and bluish colors, with fire-like reflections, somewhat irised on turning.

A bluish transparent opal, with yellowish reflections, occurs in a felsitic rick near Idaho Springs, which has been used to considerable extent in jewelry.

HYALITE. Muller's Glass.

An opal usually clear as glass and colorless, constituting globular concretions, and also crusts with a globular, reniform, botryoidal, or stalactitic surface.

Occurs near the head of Cache la Poudre Creek, and on the La Garita, in pearly crusts, on lava.

WOOD OPAL.

Wood petrified by opal of various shades of color.

Abundant along the sources of Cherry, Kiowa and Bijou Creeks, at Happy Canon, and in Middle Park, much of which is agate-like in structure, and of a diversity of colors.

BERYL.

Hexagonal; usually in long, stout six-sided prisms, without regular terminations; occasionally coarse columnar. Lustre vitreous, sometimes resinous. Color bright green (when it is known as *emerald*), pale green, passing into light blue (*aquamarine*), yellow and white. Transparent–subtranslucent. II. 7·5–8. G. 2·63–2·76. Comp., silica 66·8, alumina 19·1, glucina 14·1.

Beryls of enormous size, but nearly opaque and of inferior color, in white quartz, on Bear Creek. Small crystals of light green color in the Floyd Hill region; also on the Platte River, at the edge of South Park. Colorless and light green transparent crystals on Mt. Antero, Chaffee County.

AQUAMARINE.

Beryls of a sea-green or pale bluish-green tint.

Beautiful transparent crystals, of bluish-green color, occur on Mt. Antero, Chaffee County. Rarely on Bear Creek.

GOTHITE. Brown Iron-Stone.

Orthorhombic; in prisms longitudinally striated, and often flattened into scales or tables parallel to the shorter diagonal. Cleavage very perfect. Also fibrous, foliated, in scales, massive, stalactitic. Lustre imperfect adamantine. Color yellowish, reddish, and blackish brown. Often blood-red by transmitted light. Streak browish-yellow, ochre-yellow. H. 5–5·5. G. 4·0–4·4 Comp., sesquioxyd of iron 89·9, water 10·1.

Abundant in the Pike's Peak region, in cavities in granite, associated with amazonstone, smoky quartz, fluor spar, etc.

BASTNASITE. *Allen & Comstock, Am. Jr. Sc., May,* 1880.

Hexagonal; occurs in six-sided prisms. Color reddish-brown. Lustre vitreous to resinous. H. 4–4·5. G. 5·18–5·20. Analysis of Colorado mineral by Allen & Comstock: oxide of cerium 40·88, oxide of lanthanum and didymium 34·95, carbonic acid 20·09, fluorine (by difference) 4·08.

Occurs as an alteration product of tysonite west of Cheyenne Mountain, El Paso County.

PTILOLITE. (*new.*) *W. Cross, Am. Jr. Sc., Aug.*, 1886.
Occurs in white tufts of short hair-like needles. Analysis by Eakins: silica 70·35, alumina 11·90, lime 3·87, potassa 2·83, soda 0·77, water 10·16.

Occurs as tufts of short hair-like needles in augite-andesite. on the north slope of Green Mountain, Jefferson County.

ARAGONITE.
Orthorhombic; primary form a right rhombic prism; occurs in hexagonal prisms, very frequently in twin crystals, also in globular, reniform, dendritic, and coralloidal shapes; sometimes fibrous and in compact masses. Color generally white, but sometimes tinged yellow, blue and green. Lustre vitreous, inclining to resinous on fractured surfaces. Translucent to transparent. Fracture subconchoidal. H. 3·5-4. G. 2·031. Comp., carbonic acid 44, lime 56.

Coralloidal and in small masses in a number of lodes on Leavenworth Mountain, Georgetown. Near Golden. One mile below Idaho·Springs is a variety very similar to the celebrated *gibralter rock*, of Spain. In South Park, two miles south-west of the Salt Works, in coarsely radiated and columnar masses. On Willow Creek, San Luis Valley, a vein about 30 feet wide, in basalt, portions of which exhibit a botryoidal structure. Coralloidal at Loma. Six miles below Del Norte, on the Rio Grande, a large deposit.

EMBOLITE. Chlorobromid of Silver.
Isometric; occurs in cubes and cubo-octahedrons; also massive, sometimes stalactitic. Lustre resinous and somewhat adamantine. Color grayish-green and asparagus-green to pistachio or yellowish-green, and yellow, often dark; becoming darker externally on exposure. H. 1–1·5. G. 5·31-5·43. Comp., varies, silver 61 to 71, bromine 7 to 33, chlorine 5 to 20.

Sparingly in the mines of Leadville.

TYSONITE. (*new.*) *Am. Jr. Sc.*, May, 1880, June, 1884.
Hexagonal; occurs in prisms with basal cleavage. Color pale wax-yellow. Streak nearly white. Lustre vitreous to resinous. H. 4·5-5. G. 6·14. Analysis by Allen & Comstock: cerium 40·16, lanthanum and didymium 30·29, fluorine 29·55.

Occurs near Pike's Peak, associated with bastnasite, both minerals being frequently found in the same crystal, the central portion of the crystal being composed of tysonite. Specimens have been found weighing several pounds. Named after S. T. Tyson, a metallurgist of Colorado.

PYRITE. Iron Pyrites. Bisulphuret of Iron.
Isometric; occurs in cubes, simple and modified; pentagonal dodecahedrons and octahedrons; crystals frequently forming aggregations; also massive and in imitative shapes. Color bronze-yellow. Lustre metallic, splendent. Brittle. H. 6–6·5. G. 4·8-5·1. Comp., iron 46·7, sulphur 53·3.

Abundant in many of the mines, and usually valuable for the gold and silver it contains. Beautiful cubical crystals in the mines of Gilpin County. Dodecahedrons and cubo-octahedrons in the Josephine mine, Geneva District, Clear Creek County.

TOURMALINE.

Rhombohedral; usually in prisms terminating in a low pyramid; crystals usually hemihedral, being often unlike at the opposite extremities, or hemimorphic, and the prisms often triangular. Sometimes massive compact; also columnar, coarse or fine, parallel or divergent. Lustre vitreous. Color black, brownish-black, bluish-black, most common; blue, green, red, and sometimes of rich shades; rarely white or colorless; some specimens red internally and green externally; and others red at one extremity, and green, blue, or black at the other. Dichroic. Transparent—opaque. Pyroelectric. Brittle. H. 7–7.5. G. 2.94–3.3. Comp., black variety, silica 33, alumina 38, lime 1 protoxyd of iron 24, soda 3, boracic acid 1.

Fine black triangular prisms, with pyramidal terminations, on Bear Creek, imbedded in white quartz. On Clear Creek, two miles below black Hawk, hexagonal prisms in feldspar. At Guy Hill. Five miles north of the Platte, on the road from Colorado Springs to Fairplay, fine crystals with terminations.

PETZITE. Telluride of Gold and Silver.

Orthorhombic. Massive; compact or fine-grained; rarely coarse-granular. Lustre metallic. Color between steel-gray and iron-black, sometimes with pavonine tarnish. Streak iron-black. Brittle. H. 2.5. G. 8.72–8.83. Comp., tellurium 35, silver 47, gold 18.

In Boulder County, associated with other tellurium ores.

POLYBASITE.

Orthorhombic; occurs in short tabular, six-sided prisms, with the bases triangularly striated parallel to alternate edges. Cleavage basal, imperfect. Also massive and disseminated. Lustre metallic. Color iron-black; in thin crystals cherry-red by transmitted light. Streak iron-black. Fracture uneven. H. 2–3. G. 6.214. Analysis of polybasite from Terrible mine, Georgetown, by Dr. Genth: sulphur 16.70, antimony 10.18, arsenic 0.78, silver 62.70, copper 9.57, iron 0.97.

In small quantity in many of the silver mines. Fine crystals in the Little Emma and Colorado Central mines, Georgetown.

TETRAHEDRITE. Gray Copper. Freibergite. Fahlerz.

Isometric; tetrahedral; occurs crystallized in modified tetrahedrons, and also in compound crystals; also massive; granular, coarse or fine. Lustre metallic. Color between light flint-gray and iron-black. Streak generally same as the color: sometimes inclining to brown and cherry-red. Opaque; sometimes subtranslucent in very thin splinters, transmitted color cherry-red. Fracture subconchoidal—uneven. Rather brittle. H. 3–4.5. G. 4.5, 5.11. Comp., varies; *freibergite*, the argentiferous variety, sulphur 21 to 30, antimony 17 to 28, copper 14 to 36, iron 1 to 7, zinc 1 to 6, silver 3 to 31.

Largely disseminated through nearly all sulphide silver ores in the state. That from the mines about Georgetown carries from 20 to 30 per cent of silver, while that from contiguous districts, 3 or 4 miles from Georgetown, carries only 1 or 2 per cent of silver. Good crystals, having pavonine tints, in the Champion lode, Geneva District, Clear Creek County.

SIDERITE. Chalybite. Spathic Iron. Carbonate of Iron.

. Rhombohedral; occurs in obtuse rhombohedrons, whose faces are often curved, in acute rhombohedrons, sometimes perfect, or having the terminal angles replaced, and in lenticular crystals; also in botryoidal and globular forms, subfibrous within, occasionally silky fibrous; often cleavable massive, with cleavage planes undulating; coarse or fine granular. Lustre vitreous, more or less pearly. Color ash-gray, yellowish-gray, greenish-gray, also brown and brownish-red, rarely green, and sometimes white. Streak white. Translucent—subtranslucent. Fracture uneven. Brittle. H. 3·5-4·5. G. 3·7–3·6. Comp., carbonic acid 37·9, protoxyd of iron 62·1.

Of frequent occurrence in the mines, both massive and crystallized.

CYANITE.

Triclinic; usually in long-thin-bladed crystals aggregated together, or penetrating the gangue; crystals sometimes short and stout. Lateral cleavage distinct. Sometimes fine fibrous. Lustre vitreous—pearly. Color blue, white, blue along the center of the blades with white margins; also gray, green, black. Streak uncolored. Translucent—transparent. H. 5–.7 G. 3·6–3·7. Comp., silica 37, alumina 63.

Occurs in the quartzite of Medicine Bow Peak.

TETRADYMITE. Telluric Bismuth.

Hexagonal; crystals often tabular. Cleavage basal, very perfect. Also massive, foliated or granular. Lustre metallic, splendent. Color pale steel-gray. Laminæ flexible. Soils paper. H. 1·5–2. G. 7·2–7·9. Comp., tellurium 48, bismuth 52.

At Gold Hill, Boulder County, associated with other tellurium ores.

XENOTIME. Phosphate of Yttria. *W. E. Hidden, Am. Jr. Sc., March,*1885.

Tetragonal; primary form a rectangular prism with a square base; occurs in square octahedrons, with perfect prismatic cleavage. Color yellowish-brown, reddish-brown, hair-brown, flesh-red, grayish-white, pale yellow. Streak pale brown, yellowish or reddish. Opaque. Fracture uneven and splintery. H. 4–5. G. 4·45–4·56. Comp., phosphoric acid 37·86, yttria 62·14.

Small chocolate-brown crystals west of Cheyenne Mountain, El Paso County, from same locality as tysonite and bastnasite.

TORBERNITE. Copper-Uranite.

Tetragonal; in square tables, often with replaced edges. Cleavage basal, highly perfect, micaceous. Lustre pearly—adamantine. Color emerald- and grass-green, and sometimes leek-, apple- and siskin-green. Streak somewhat paler than the color. Transparent—subtranslucent. Sectile. Laminæ brittle and not flexible. H. 2–2·5. G. 3·4–3·6. Comp., phosphoric acid 15·7, oxyd of uranium 62·7, lime 6·1, water 15·5.

Sparingly in the Peabody lode, Georgetown.

BROMYRITE. Bromic Silver. Bromid of Silver.

Isometric; occurs in cubes, octahedrons and cubo-octahedrons; but generally in small concretions. Lustre splendent. Color when pure, bright yellow to amber colored, slightly greenish; often grass-green or olive-green externally. Sectile. H. 2–3. G. 5·8–6. Comp., bromine 42·6, silver 57·4.

Occasionally found in the mines at Leadville.

MINERAL COAL.

The distinguishing characters of mineral coal are as follows: Compact massive, without crystalline structure or cleavage; sometimes breaking with a degree of regularity, but from a jointed rather than a cleavage structure. Sometimes laminated; often faintly and delicately banded, successive layers differing slightly in lustre. Lustre dull to brilliant, and either earthy, resinous or submetallic. Color black, grayish-black, brownish-black, and occasionally iridescent; sometimes dark brown. Opaque. Fracture conchoidal—uneven. Brittle; rarely somewhat sectile. Infusible to subfusible; but often becoming a soft, pliant, or paste-like mass when heated. On distillation most kinds afford more or less of oily and tarry substances, which are mixtures of hydrocarbons and paraffin. Hardness 0·5—2·5. Gravity 1—1·80.

The varieties depend partly (1) on the amount of the volatile ingredients afforded on destructive distillation; or (2) on the nature of these volatile compounds, for ingredients of similar composition may differ widely in volatility; (3) structure, lustre, and other physical characters. The following varieties are found in Colorado:

BROWN COAL. Lignite.

Sometimes pitch-black, but often rather dull and brownish-black. Fragile compared with anthracite. Occasionally somewhat lamellar in structure.

The lignite coal area of Colorado extends from St. Vrains on the north to the Raton Mountains on the south, about 220 miles in length, and varying from 20 to 35 miles in breadth along the eastern base of the mountains. Large portions of this field have been swept away by floods resulting from melting glaciers. The principal developments of this vast field have been made in the vicinity of Boulder, Erie, Loveland, Golden, Colorado Springs, Coal Creek, Canon City, Walsenburg and Trinidad. Lignite also exists in North Park and at various points on the divide between North and Middle Parks. The following analyses have been made:

	Water.	Volatile Matter.	Fixed Carbon.	Ash.	Sulphur.
Coal Hill, North Park	18·35	32·20	41·90	6·45	1.10
do do	8·82	41·56	41·17	6·00	2·45
Red Hill, do	15·20	33 30	48·00	3·50	
do do	10·50	37·30	50·30	1·90	
No. 1, do	13·20	9·30	72·00	5·50	
Erie Mine. Weld County	14·80	34·50	47·30	3·40	
Marshal Mine, Boulder County	12·00	26·00	59·20	2·80	
Murphy Mine, Jefferson County	13·83	35·88	44·40	5·85	
Golden, do do	13·43	37·15	45·57	3·85	
Franceville, El Paso County	12·90	29·10	46·00	12·00	
Cutler banks, Uncompahgre County	7·26	43·42	41·72	7·60	

JET.

Jet is a black variety of brown coal, compact in texture, and taking a good polish, whence its use in jewelry.

Occurs in seams, from one-half to six inches in width, in the shale about Canon City and Little Fountain Creek.

BITUMINOUS COAL.

Bituminous coals have the common characteristic of burning in the fire with a yellow, smoky flame, and giving out on distillation hydrocarbon oils or tar. They have usually a bright, pitchy, greasy lustre. The caking variety becomes pasty or semi-viscid in the fire. The softening takes place at the temperature of incipient decomposition, and is attended with the escape of bubbles of gas. On increasing the heat, the volatile products which result from the ultimate decomposition of the softened mass are driven off, and a coherent, grayish-black, cellular, or fritted mass (coke) is left. A caking coal will lose its caking quality if kept heated for 2 or 3 hours at 300°C., and sometimes on mere exposure for a time to the air. Non-caking coal is like the preceding in all external characters, and often in ultimate composition; but burning freely without softening. The coke is not a proper coke, being in powder, or of the form of the original coal.

The Trinidad group of mines, embracing those in the immediate vicinity of Trinidad with those of ElMoro and Starkville, and the mines at Crested Butte, are the most important of this class in the State. Durango, Aspen and Rico are other localities of its occurrence.

	Water.	Volatile Matter.	Fixed Carbon.	Ash.
Engle Mine, El Moro, Las Animas County......	·95	29·82	56·41	12·82
do do do do	·26	29·66	65·76	4·32
Cucharas, Huerfano County.........................	3·23	40·93	49·54	6·30
Crested Butte, Gunnison County....................	3·70	30·97	61·07	4·47
do do	·44	24·17	72·30	3·09
SEMI-BITUMINOUS.				
Col. Coal and Iron Co., Park County..............	5·03	35·85	50·97	8·15
Oak Creek, do	6·72	34·76	52·70	5·82

Semi-bituminous coal is found principally in the mountainous part of the state, The mines at Como on the South Park railway are the most productive of this class. A good vein has also been opened up on Mt. Carbon, 17 miles north-west of Gunnison City. The coal from Walsenburg is by many regarded as belonging to this class.

BYERITE.

Color jet-black; vandyke-brown when powdered. Has the appearance of bituminous coal. Does not slack or break to pieces. In composition much like albertite and torbanite. G. 1·323. Analysis by E. J. Mallett:

Water .... 6·02
Volatile matter .... 39·95
Fixed carbon and ash................. 54·03—100·00

The locality of this new coal species is Middle Park. It is a caking bituminous coal but yields an imperfect coke. In comparison with English cannel coal it gives thirty per cent more gas of a high illuminating power. It is named after Mr. W. N. Byers.

NATIVE COKE.

More compact than artificial coke, and sometimes affording considerable bitumen.

The origin of native coke is attributed to the action of a

trap eruption on bituminous coal. It has been found near Crested Butte where a dike of lava has intruded the coal strata. It has also been found at Trinidad.

ANTHRACITE.

Lustre bright, often submetallic, iron-black, and frequently iridescent. Fracture conchoidal. Volatile matter after drying 3 to 6 per cent. Burns with a feeble flame of a pale color. H. 2-2·5.

Anthracite coal appears to be confined mainly to the coal basins in the Elk Mountains, and extends over a large scope of country. These mountains are composed of lofty peaks of eruptive rock, and are separated from the plains by three parallel ranges, the Colorado or Front Range, the Park Range and the Saguache Range, separated from one another by trough-like valleys and parks. The coal fields are in the same geological horizon as that of the plains, and the coal is much the same, metamorphosed or changed by heat into anthracite, and on the outer edges of the basins graduates into lignite. On Anthracite Mesa, near Crested Butte, the veins are from two to ten feet thick. Analyses show

		Water.	Volatile Matter.	Fixed Carbon.	Ash.
Crested Butte, Gunnison County		1·20	5·16	90·24	3·40
Anthracite Creek,	do	2·00	2·50	91 90	3·60
Anthracite Mesa,	do		5·17	90·85	4·18
Rock Creek,	do		7·40	88·90	3·68
SEMI-ANTHRACITE.					
W. & D. Trust Co., Crested Butte		4·00	14·00	74·00	8·00
Uncompahgre Canon		1·86	10·70	77·32	10·12

PETROLEUM. Naptha. Mineral Oil. Kerosene.

A fluid bitumen of dark color, which oozes from certain rocks and becomes solid on exposure, forming asphaltum. Com., carbon 82·2, hydrogen 14·8.

Oozes from sandstone near Morrison. Six oil wells are in operation in the neighborhood of Canon City and Florence, in Fremont county. The Oil Creek well, which was sunk in 1862, turned out about 3,000 gallons of oil.

ASPHALTUM. Bitumen. Mineral Pitch.

Amorphous. Lustre like that of black pitch. Color brownish-black and black. Odor bituminous. Melts ordinarily at 90° to 100° C., and burns with a bright flame. Soluble mostly or wholly in oil of turpentine, and partly or wholly in ether; commonly partly in alcohol. The more solid kinds grade into the pittasphalts or mineral tar, and through these is a gradation to petroleum. Comp., carbon 75·0, hydrogen 9·5, oxygen 15·5.

Fluid bitumen oozes from sandstone near Morrison and forms an earthy asphaltum. In the White River region and near Canon City.

MINERAL RESIN.

Mineral resin occurs in the coal in scales, and small masses have been found in the soil on the Table Mountains, but the species have not been determined. Prof. Wm. P. Headden mentions the occurrence of a substance in coal from the Mitchell mine, Erie, which he believes to be *napthaline.* It occurs in microscopic, tabular crystals, which are thin, colorless and transparent.

TURGITE.

Compact fibrous and divergent, to massive; often botryoidal and stalactitic like limonite. Also earthy, as red ochre. Lustre submetallic and somewhat satin-like in the direction of the fibrous structure; also dull earthy. Color reddish-black, to dark red; bright-red when earthy; botryoidal surface often lustrous, like much limonite. Opaque. H· 5–6. G. 3·56–3·74. Comp., Sesquioxyd of iron 94·7. water 5·3.

Occurs in small botryoidal masses, of reddish-black color, and highly polished surfaces, at Florissant, associated with gothite, amazonstone, smoky quartz, etc.

ALMANDITE. Iron-Alumina Garnet. Precious or Oriental Garnet.

Isometric; occurs in dodecahedrons and trapezohedrons, both forms sometimes variously modified; rarely in octahedrons. Cleavage parallel to the faces of the dodecahedron. Also found massive. Color fine deep-red and transparent, and then called *precious garnet;* also brownish-red, and translucent or subtranslucent. Streak white. Lustre vitreous—resinous. Fracture subconchoidal, uneven. H. 6·5–7·5. G. 4·13. Comp., silica 36·1, alumina 20·6, protoxyd of iron 43·3.

Large and perfect dodecahedrons, some of them measuring six inches in diameter, occur in chlorite near Salida, Chaffee County. Several thousand specimens have been taken out of a single "pocket." They are usually coated with chlorite, and the surfaces of the garnets are decomposed, the decemposition product forming *aphrosiderite.* The color of the garnet is brownish-red. Analysis by S. L. Penfield and F. L. Sperry gave the following:

Silica	37·61
Alumina	22·70
Protoxyd of Iron	33·83
Protoxyd of Manganese	1·12
Magnesia	3·61
Lime	1·44
	100·31

Beautiful dark red garnets, which can probably be classed as precious garnets, are abundant in the placers at Fairplay and Breckenridge. The are usually quite small and of no value as gems.

APHROSIDERITE.

A soft ferruginous chlorite, of a dark olive-green color, scaly massive in structure; the scales minute, transparent, and hexagonal.

Occurs as an alteration product on the garnets found at Salida, forming a coating of light green color. Analysis by Penfield and Sperry gave:

Silica	28·20
Alumina	22·31
Protoxyd of Iron	19·11
Protoxyd of Manganese	17·68
Lime	·48
Soda	·72
Potassa	1·03
Water	10·90

FAYALITE. Iron Chrysolite. Anhydrous Silicate of Iron.
Massive, crystalline. Cleavage in two directions at right angles to one another. Lustre metalloid, somewhat resinous in the fracture. Color black; sometimes iridescent. Opaque. Fracture imperfectly conchoidal. Attractable by the magnet. H. 6·5. G. 4–4·14. Comp., silica 29·5, protoxyd of iron 70·5.

Occurs in quite large masses in Cheyenne Canon, El Paso County.

FIBROLITE.
Monoclinic; crystals commonly long and slender. Cleavage brilliant and perfect. Also occurs fibrous or columnar massive, sometimes radiating. Lustre vitreous, subadamantine. Color hair-brown, grayish-brown, grayish-white, grayish-green, pale olive-green. Streak uncolored. H. 6–7. G. 3·2. Comp., silica 36·8, alumina 63·2.

Occurs in small fibrous masses, of grayish-green color, on Sherman Mountain, near Georgetown.

SANIDIN. Glassy Feldspar. *Whitman Cross, Am. Jr. Sc., Feb.,* 1884.
A variety of orthoclase (p.12) occurring in transparent glassy crystals, mostly tabular, in eruptive rock.

Mr. Cross describes the occurrence of sanidin in the nevadite of Chalk Mountain, and rhyolite of Ragged Mountain, which possesses a satin lustre. Smoky quartz and topaz are associates of the sanidin in the nevadite. Also occurs in rhyolite at Nathrop, with the same associate minerals.

DOLOMITE. Pearl Spar. Brown Spar. Magnesian Limestone.
Rhombohedral; crystals often having curved faces. Cleavage perfect parallel to the primary faces. Often granular and massive, constituting extensive beds. Color white or tinged with yellow, red, green, brown and sometimes black. Lustre vitreous or a little pearly. Nearly transparent to translucent. Brittle. H. 3·5–4. G. 2·8–2·9. Comp., carbonate of lime 54·35, carbonate of magnesia 45·65.

Occurs in extensive beds in various portions of the State, and in small quantities in the mines, frequently crystallized.

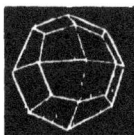

APPENDIX.

ENSTATITE. Bronzite. Protobastite. *W. Cross*, Bulletin No. 1, U. S. Geological Survey.

Orthorhombic; primary form an oblique four-sided prism with a very distinct cleavage parallel to the lateral planes; sometimes a fibrous appearance on the cleavage surface; also occurs massive and lamellar. Lustre a little pearly on the cleavage-surfaces to vitreous; often metalloidal in the bronzite variety. Color grayish-white, yellowish-white, greenish-white, to olive-green and brown. Streak uncolored, grayish. Double refraction positive. H. 5·5. G. 3·1—3·3. Composition:

Silica	60
Magnesia	40
Sometimes iron, alumina, etc.	——
	100

Whitman Cross gives the occurrence of an enstatite-bearing diabase in a narrow dike of dark rock at Morrison.

LEPIDOMELANE. Iron Potash Mica.

Hexagonal; occurs in small six-sided tables, or an aggregation of minute opaque scales, with perfect basal cleavage. Lustre adamantine, inclining to vitreous, pearly. Color black with occasionally a leek-green reflection. Streak grayish-green. Opaque or translucent in very thin laminæ. Somewhat brittle, or but little elastic. H. 3. G. 3. Composition:

Silica	37·40
Alumina	11·60
Peroxide of Iron	27·66
Protoxide of Iron	12·43
Magnesia	0·26
Potash	9·20
Water	0·60
	99·69

According to Hague, some of the mica in the granite of the eastern Colorado range is lepidomelane.

RHODONITE. Manganese Spar. Bisilicate of Manganese.

Triclinic, and like pyroxene, with a cleavage in three directions, two of which are perpendicular to each other. Usually massive. Lustre vitreous. Color light brownish-red, flesh-red, sometimes greenish or yellowish, when impure; often black outside from exposure. Streak white. Transparent—opaque. Fracture conchoidal—uneven. Very tough when massive. H. 5·5—6·5. G. 3·4—3·68 Composition:

Silica	45·9
Protoxide of Manganese	51·4
	100·0

Occurs as a gangue of light rose color in the San Juan mines.

ALAMANDITE (p. 48).—New localities in Chaffee County mentioned by W. B. Smith. Dodecahedral crystals in chloritic schist from Longfellow gulch. Crystals in limestone, having the dodecahedron and trapezohedron about equally developed, from Calumet.

CINNABAR is said to be found near Durango, but needs verification.

ENARGITE (p. 13).—Occurs massive and in slender crystals in the Missouri mine, Hall Valley, Park County. Small masses in the Centennial mine, Georgetown. In considerable quantity·in the Forest Queen mine, Gunnison County, and in the mines of Red Mountain district, San Juan.

GRAPHITE (p. 27).—A large deposit a few miles south of the El Moro coal banks, Las Animas County.

NATIVE SULPHUR (p. 6).—Abundant in the Spirit mine at Red Cliff.

PETROLEUM (p. 47).—Has been found on Green River.

TENNANTITE (p. 8).—In large quantities in the Red Mountain mines, San Juan.

TORBERNITE (p. 44).—Occurs on Lyden Creek (Berthoud).

MINIUM. Oxide of Lead.
Occurs pulverulent, occasionally exhibiting, under the microscope, crystalline scales. Lustre faint greasy, or dull. Color vivid red, mixed with yellow. Streak orange-yellow. Opaque. H. 2—3. G. 4·6. Composition:

Oxygen	9·34
Lead	90·66
	100·00

Prof. Emmons mentions its occurrence in some of the mines at Leadville.

HEULANDITE.
Monoclinic; primary form a right rhombic prism. Occurs in attached crystals and in layers and granular masses, frequently in a globular form. Lustre vitreous; pearly on planes of cleavage. Color various shades of white, passing into red, gray and brown. Streak white. Transparent—subtranslucent. Fracture subconchoidal, uneven. Brittle. H. 3·5—4. G. 2·2. Composition:

Silica	59·1
Alumina	16·9
Lime	9·2
Water	14·8

Occurs in small white glassy crystals in the augite-andesite on Green Mountain, Jefferson County.

HEULANDITE.

Monoclinic; primary form a right rhombic prism. Occurs in attached crystals and in layers and granular masses, frequently in a globular form. Lustre vitreous; pearly on planes of cleavage. Color various shades of white, passing into red, gray and brown. Streak white. Transparent—subtranslucent. Fracture subconchoidal, uneven. Brittle. H. 3·5—4. G. 2·2. Composition:

Silica	59·1
Alumina	16·9
Lime	9·2
Water	14·8
	100·0

Occurs in small white glassy crystals in the augite-andesite on Green Mountain, Jefferson County.

STRONTIANITE. Carbonate of Strontian.

Orthorhombic; primary form a right rhombic prism; occurs crystallized in hexahedral prisms, which are modified on the edges, or terminated by pyramids; often acicular and in divergent groups; also occurs in columnar-globular forms, fibrous and granular. Lustre vitreous, inclining to resinous on uneven faces of fracture. Color pale asparagus-green, apple-green, also white, gray, yellow, and yellowish-brown. Streak white. Transparent—translucent. Fracture uneven. Brittle. H. 3·5—4. G. 3·605—3·713. Composition:

Carbonic Acid	29·8
Strontia	70·2
	100·0

Occurs in the shape of small balls on celestite from the the Garden of the Gods.

STROMEYERITE. Sulphuret of Silver and Copper.

Orthorhombic; isomorphous with copper glance. Usually massive. Lustre metallic. Color dark steel-gray. Streak shining. Brittle. Fracture subconchoidal. H. 2·5—3. G. 6·2—6.3. Composition, silver, sulphur and copper in variable proportions. Partial analyses of stromeyerite from Colorado mines returned the following :

	Ag.	Cu.	S.	Fe.	
Little Giant mine, Clear Creek Co	50·00	21·00	16·60	3·40	—W.C. Minger.
" " " " " "	48·00	35·00	16·00		—R. Neuman.
Grant mine, Boulder County	38·00	29·00			—F. Graham.
Winning Card mine, Summit Co.	49·00	36·00	16·00		—R. Neuman.

Occurs in small masses and needle crystals coated with chalcopyrite in the Little Giant mine, Clear Creek County. In considerable quanty in the Grant mine, Boulder County, and in the Winning Card and Black Prince mines, Summit County. A variety carrying 1½ p.c. silver is found in small quantities in a lode on Saxon Mountain, Georgetown.

RHODOCHROSITE (p. 10.)—Crystallized and fibrous in the Queen of the West mine, near Gray's Peak, associated with alabandite.

RUTILE. Oxide of Titanium.

Tetragonal; occurs in four or eight-sided prisms, terminated by pyramids, either single or geniculated, and often striated longitudinally; also in reticulated masses formed by acicular and capillary macled crystals; also massive and imbedded. Structure lamallar. Color reddish-brown, passing into red, sometimes yellowish, bluish, violet, black, rarely grass-green. Streak pale brown. Lustre metallic adamantine. Fracture subconchoidal, uneven. Subtransparent—opaque. Brittle. H. 6—6·5. G. 4·18—4·25. Composition: Titanic Acid:

Oxygen	39
Titanium	61
	100

A small quartz pebble containing red rutile was found in the Platte River at the foot of the mountains. It is in the collection of Whitman Cross.

COSALITE (p. 13).—Analyses: I from the Alaska mine, Ouray County, by Genth. II from Ouray County, by Konig:

	I.	II.
Bismuth	44·95	43·54
Lead	28·10	26·77
Silver	1·44	1·35
Copper	8·00	8·22
Zinc	0·24	......
Antimony	0·51	......
Arsenic	0·04	......
Selenium	trace	......
Sulphur	16·80	16·54
	100·08	96·42

BEEGERITE (p. 8).—Analyses: I from Ouray county, by Konig. II from the Treasure Vault mine, Geneva District, Clear Creek County, by Genth.

	I.	II.
Bismuth	19·35	19·81
Lead	45·87	50·16
Silver	9·98	15·40
Sulphur	13·37	14·59
	88·57	99·96

ANNABERGITE (p. 37).—Analysis of annabergite from the Gem mine, Silver Cliff, by Genth:

Arsenic Acid	36·64
Oxide of Nickel	32·64
Oxide of Cobalt	0·50
Lime	3·51
Magnesia	3·74
Water	23·94
	100·97

CUPRITE (p. 40).—Very pure in veins near Pine Grove, Platte Canon, associated with chrysocolla.

WITTICHENITE. Cupreous Sulphuret of Bismuth.
Orthorhombic; occurs massive and disseminated; also coarse columnar; or an aggregate of imperfect prisms. Cleavage in one vertical direction. Color steel-gray, tin-white, tarnishing pale lead-gray. Streak black. H. 3·5. G. 5. Composition:

Sulphur	19·44
Bismuth	42·11
Copper	38·45
	100·00

According to T. B. Comstock wittichenite is of common occurrence in the mines of San Juan county, and is rich in silver.

AIKINITE. Acicular Bismuth. Needle Ore.
Orthorhombic; occurs in imbedded acicular four or six-sided prisms, indistinctly terminated, and striated longitudinally. Also massive. Lustre metallic. Color blackish lead gray, with a pale copper-red tarnish. Fracture uneven. H. 2—2·5. G. 6·1—6·8. Composition:

Sulphur	16·7
Bismuth	36·2
Lead	36·1
Copper	11·0
	100·0

Occurs in the mines of San Juan County in beautiful crystals, and also in the Gladiator mine, Hinsdale county, carrying about 5 per cent of silver (T. B. Comstock).

YTTROTANTALITE.
Orthorhombic; crystals often tabular; also massive and amorphous. Lustre submetallic to vitreous and greasy. Color black, brown, brownish-yellow, straw-yellow. Streak gray to colorless. Opaque to subtranslucent. Fracture small conchoidal to granular. H. 5—5·5. G. 5·4—5·9. Composition:

Tantalic Acid	62·5
Yttria	22·6
Lime	5·2
Protoxide of Iron	3·4
Protoxide of Uranium	6·3
	100·0

W. B. Smith mentions the occurrence of a black variety, in small masses, at Devil's Head Mountain, Douglas County.

WAD. Bog Manganese.
Occurs in amorphous and reniform masses, either earthy or compact, and sometimes incrusting or as stains. Often loosely aggregated, and feeling very light to the hands. Color dull black, bluish or brownish-black. H. 0·5. G. 3—4·26. Composition, oxide of manganese, but mixed with other ingredients.

Occurs in small quantities in many localities throughout the state, frequently staining the rocks black at the surface of veins, and sometimes forming beautiful dendritic specimens, called "forest rock." Quite a large deposit near the springs on Snake River, Summit County.

HYPERSTHENE. Metalloidal Diallage. Labrador Hornblende.
Orthorhombic; occurs in foliated masses or imbedded in rocks. Lustre somewhat pearly on a cleavage face, sometimes metalloidal. Color dark brownish-green, grayish-black, pinchbeck-brown. Streak grayish, brownish-gray. Translucent to nearly opaque. Brittle. H. 5—6. G. 3·392. Analyses of hypersthene from Buffalo Peaks, Mosquito Range, by Hillebrand:

Silica	51·703	51·157	50·043
Alumina	1·720	2·154	2·906
Sesquioxide of Iron	0·304	....	....
Protoxide of Iron	17·995	18·360	17·812
Protoxide of Manganese	0·363	0·363	0·120
Lime	2·873	3·812	6·696
Magnesia	25·091	24·251	21·744
Soda	....	....	0·274
	100·049	100·097	99·595

One of the essential constituents of the andesite and tufa rocks of Buffalo Peaks, Mosquito Range.

PYROXENE.
Monoclinic; primary form an oblique rhombic prism; generally occurs in short, thick crystals, and often in twins; also amorphous, coarsely laminar, granular and fibrous, fibres often fine and long. Lustre vitreous inclining to resinous, sometimes pearly. Color green of various shades, verging on one side to white or grayish-white, and on the other to brown or black. Streak white to gray and grayish-green. Fracture conchoidal, uneven. Brittle. H. 5—6. G. 3·23—3·5. Composition: Bisilicate of various bases, the bases being lime, magnesia, protoxide of iron, protoxide of manganese, and sometimes potash and soda.

Next to the feldspars, pyroxene is the most universal constituent of the igneous rocks of the mountains.

PYROLUSITE. Gray Oxide of Manganese.
Orthorhombic; occurs crystallized, but generally in botryoidal and reniform masses, with a radiating, fibrous or columnar structure; or in granular masses. Often soils. Lustre metallic. Color iron-black, dark steelgray, sometimes bluish. Streak black or bluish-black, sometimes submetallic. Rather brittle. H. 2—5·5. G. 4·82. Composition:

Manganese	63·3
Oxigen	36·7
	100·00

Peale mentions its occurrence on Silver Heels Mountain, near Fairplay, and in the San Juan country.

MANGANITE.
Orthorhombic; primary form a right rhombic prism; occasionally hemihedral; occurs in columnar crystals striated vertically; also fibrous and massive, or radiating; sometimes granular. Lustre submetallic. Color dark steel-gray, iron-black. Streak reddish-brown, sometimes nearly black. Opaque; minute splinters sometimes brown by transmitted light. Fracture uneven. H. 4. G. 4·2—4·4. Composition:

Sesquioxide of Manganese	89·8
Water	10·2
	100·0

W. B. Smith reports its occurrence in small masses in the region of Devil's Head Mountain, Douglas County.

ARGENTOBISMUTITE.

F. A. Genth gives the following analysis of argentobismutite from Lake City, Hinsdale County:

Sulphur	16.66
Bismuth	52.89
Silver	26.39
Lead	4.06—100.00

BOURNONITE. Antimonial Lead Ore.

Orthorhombic; occurs crystallized in modified rectangular prisms, often cruciform or compounded into shapes like a cog-wheel. Also massive, granular compact. Lustre metallic. Color and streak steel-gray, inclining to blackish lead gray or iron black. Opaque. Fracture conchoidal or uneven. Brittle. H. 2.5—3. G. 5.7—5.9. Composition:

Sulphur	19.7
Antimony	25.0
Lead	42.4
Copper	12.9—100.00

M. C. Ihlseng mentions the common occurrence of this mineral in the lodes of Bear and Anvil Mountains, San Juan County. Its associate mineral is galena.

BYSSOLITE.

Asbestus (p. 30) of an olive-green color. Fibres usually coarse and stiff.

Needle-like fibres an inch or more in length in clear quartz crystals near Calumet, Chaffee County.

ELECTRUM.

A natural alloy of gold and silver in the proportion of two of gold to one of silver, or,

Gold	65
Silver	35—100

Color pale yellow to yellowish-white. Known crystals tabular, and imperfect cubes. G. 14 to 17.

According to P. H. VanDiest the ore in the Nelly mine, San Miguel County, is crystallized electrum.

MELANTERITE. Copperas. Green Vitriol. Sulphate of Iron.

Monoclinic; in acute rhombic prisms. Occurs massive, pulverulent, botryoidal, reniform, stalactitic and crystallized. Lustre vitreous. Color various shades of green, passing into white; becoming yellowish-white on exposure. Subtransparent to translucent. Taste sweetish, astringent, and metallic. Fracture conchoidal. Brittle. H. 2. G. 1.832. Composition:

Sulphuric Acid	28.8
Protoxyd of Iron	25.9
Water	45.3—100.0

Proceeds from the decomposition of pyrite in the Black Iron and other mines around Red Cliff; clear green crystals and large masses. Crystals and small masses in the Pennsylvania mine, Peru District, Summit County.

NATIVE CHARCOAL.

F. F. Chisholm mentions the occurrence of native charcoal in considerable quantities in the anthracite on the head of Elk Head Creek in Routt County. According to Chisolm it is of common occurrence in Colorado lignites. Charcoal is found at various depths in the Bassick mine, Custer County.

ALUNITE. Alumstone.

Rhombohedral; crystals modifications of an obtuse rhomboid. Massive, having a fibrous, granular, or impalpable texture. Lustre of crystals vitreous, or a little pearly on the basal plane. Color white, grayish or reddish. Streak white. Transparent—subtranslucent. Fracture, flat conchoidal, uneven; of massive varieties splintery; sometimes earthy. Brittle. H. 3˙5—4. G. 2.58—2˙752. Composition:

Sulphuric Acid	38.53
Alumina	37.13
Potash	11.34
Water	13.00—100.00

S. F. Emmons mentions its occurrence in the Iron mine at Leadville. Probably occurs at a number of places along the foot-hills with native alum.

SIDEROPHYLLITE. (New). Black Mica. *H. C. Lewis*, Proc. Ac. Nat. Sc Phila., 1880.

Color black; by transmitted light chrome green. Brittle. Analysis:

Silica	36.68
Alumina	20.41
Sesquioxyd of Iron	1.55
Protoxyd of Iron	25.50
Protoxyd of Manganese	2.10
Magnesia	1.14
Lime	0.81
Soda	1.09
Lithia	0.37
Potassa	9.20
Water	1.01—99.86

Found in the Pike's Peak region.

TELASPYRINE. (New). Telluride of Iron. *C. U. Shepard*, Contributions to Mineralogy, 1877.

Color silver-white. Analysis by Edmond Fuchs:

Gold	5.228
Silver	4.198
Tellurium	19.650
Iron	70.624
Sulphur	trace—99.700

Occurs in the Grand View and Phil Sheridan mines, Sunshine, Boulder County.

URACONITE. Sulphate of Uranium.

Amorphous, earthy, or scaly, and of fine lemon-yellow color, or orange. Composition:

Sulphuric Acid	10.16
Sesquioxyd of Uranium	66.05
Sesquioxyd of Iron	0.86
Lime	2.62
Water	20.06—99.76

In the Wood lode, Leavenworth Gulch, near Central City, with pitchblende.

STIBNITE (p. 23).—Large veins of sulphide of antimony between Troublesome Creek and Lost Lake in Grand County. Float masses of stibnite and quartz are found two or more feet in thickness.

ANGLESITE (p. 7).—Occurs as the oxidation product of galena in the mines around Red Cliff. It is abundant in the form of minute crystals, or "sand."—*Emmons.*

ANTHRACITE (p. 47).—True anthracite is found on the head of Elk Head Creek in Routt Co. F. F. Chisolm, who examined the field, says there are three veins, the upper four feet thick, the second three and one half feet and the third one foot in thickness. Analyses from the two upper veins:

Sulphur	0.79	0.44
Moisture	1.02	2.50
Volatile matter	9.66	3.20
Fixed Carbon	83.50	88.20
Ash	5.82	6.10

CALAMINE v. 27).—Abundant and crystallized in the Sunnyside mine, San Juan County.—*Ihlseng.*

GADOLINITE.—Massive in the upper canon of the South Platte River.

JAROSITE (p. 7).—Minute but perfect crystals in the Black Iron mine near Red Cliff.—*Emmons.*

LIMONITE (p. 33).—Pseudomorphs after pyrite in the Nelly mine, San Miguel County.

MOLYBDENITE (p. 25).—Common in the lodes on Cement Creek and near Chattanooga, San Juan County.—*Ihlseng.*

PHENACITE (p. 7).—Occurs in prismatic crystals on aquamarine on Mt. Antero in Chaffee County, and is thought to be the decomposition product of beryl.

RUTILE (p. 58).—W. B. Smith found black ferriferous rutile on St. Peter's Dome, El Paso County.

GARNET (p. 33.)—Dr. Konig publishes the following analysis of a massive titaniferous garnet from southern Colorado, and comments on its relation to schorlomite:

Silica	30·71
Titanic Acid	8·11
Alumina	2·20
Sesquioxide of Iron	23·20
Lime	31·40
Magnesia	1·22
Protoxide of Manganese	0·46—99.98

PYRITE (p. 42)—Octahedral crystals on Mount Richard Owen, Anthracite Creek, Gunnison County.

CHRYSOCOLLA (p. 29).—Occurs near Pine Grove, in Platte Canon, associated with red oxide of copper.

BISMUTHINITE (p. 9).—Occurs in the mines of San Juan County.

CORUNDUM. Emery. Blue, Sapphire. Red, Ruby.

Rhombohedral; commonly occurs crystallized in six-sided prisms; also in obtuse and acute hexahedral pyramids; cleavage basal, sometimes perfect, but interrupted. Also massive, granular or impalpable. Lustre vitreous, sometimes pearly in the basal planes, and occasionally exhibiting a bright opalescent star of six rays in the direction of the axis (asteriated sapphire). Color blue, red, yellow, brown, gray and nearly white. Transparent to translucent. Fracture conchoidal. Exceedingly tough when compact. H. 9. G. 3.909—4.19. Composition, pure alumina:

Oxygen	46.6
Aluminum	53.4—100.0

W. B. Smith mentions the occurrence of a corundum shist in Chaffee County which contains 35 per cent of corundum. It occurs in thin flat crystals, and many of them are true sapphires.

SAPPHIRE. Blue Corundum.

The blue variety of corundum.

Thin flat crystals, the largest not more than 5^{mm} in diameter, are quite abundant in the corundum shist from Chaffee County. The crystals are transparent and true sapphires.

PLAGIOCLASE. Triclinic Feldspar.

The name for the group of triclinic feldspars, the two prominent cleavage directions of which are oblique to one another.

The most abundant constituent of the andesite of Buffalo Peaks.—*W. Cross.*

THENARDITE. Anhydrous Sulphate of Soda.

Orthorhombic: cleavage basal; primary form a right rhombic prism. Occurs in rhombic octahedrons, simple or modified on the summit, aggregated in crusts and druses. Color white. Translucent or pellucid. Lustre vitreous. Effloresces and becomes covered on the surface with a white powder on exposure to the air. Taste saline. Wholly soluble in distilled water. Refracts doubly. H. 2.5. G. 2.6—2.73. Analysis:

Water and organic matter	19.10
Silica	5.20
Alumina and oxide of iron	14.40
Chlorine	2.40
Sulphuric Acid	25.80
Lime	8.80
Magnesia	3.90
Soda	19.10—98.70

Occurs in quantity at Burdsall's Lake near Morrison.

ZIRCON (p. 58).—Small crystals in the porphyrite of Ten-Mile District, Summit County.—*W. Cross.*

POLYBASITE (p. 43).—The distinguishing silver ore of Upper San Miguel District and Marshal Basin, San Miguel County.—*Ihlseng.*

ASBESTUS (p. 30).—Beautiful snow-white asbestus is found near Jasper, Rio Grande County. Asbestus is also found in float rock on Arapahoe Peak.

DYSCRASITE. Antimonial Silver.
Orthorhombic; occurs in hexagonal prisms and stellate forms; also mass-ive, disseminated, or in grains. Color and streak between silver-white and tin-white, often tarnished yellow or reddish. Lustre metallic. Easily frangible. Soft and slightly malleable. H. 3.5—4. G. 9.44—9.82. Comp.:

Silver 78
Antimony .... 22—100

Hills and Endlich mention the occurrence of dyscrasite in the mines of Poughkeepsie Gulch, San Juan county.

HYDROPHANE. Variety of Opal.
Massive, amorphous. Lustre vitreous, frequently subvitreous. Translu-cent. Color whitish, or light colored. Adheres to the tongue and becomes more translucent or transparent in water. Composition:

Silica 93.00
Alumina 2.00
Water .... 5.00—100.00

A white and nearly opaque variety of hydrophane, in rounded lumps, with a white chalky or glazed coating, is found in the State, but the locality is a secret. When water is dropped upon it, it first becomes very white and chalky, and then gradually perfectly transparent. This property is developed so strikingly that the name "Magic Stone" has been proposed for it, and its use suggested in rings, lockets, charms, etc., to conceal photographs, hair or other objects which the wearer wishes to reveal only when his caprice dictates (*Kunz*).

RIONITE. Bismuth-Tetrahedrite.
Has a conchoidal fracture, an iron-black color, black streak and greasy-metallic lustre. An analysis of the Switzerland mineral gave:

Sulphur .... 29·10
Arsenic .... 11·44
Antimony .... 2·19
Bismuth........................... 13·07
Copper .... 37·52
Silver .... 0·04
Iron ·.............................. 6·51
Cobalt .... 1·20—101·07

Abundant in the Yankee Girl mine, Ouray county, carrying about two per cent of silver.

ZORGITE. Seleniuret of Lead and Copper.
Massive, granular. Lustre metallic. Color dark or light lead-gray, some-times inclining to reddish, and often with a brass-yellow or blue tarnish. Streak darker. Brittle. H. 2·5. G. 7—7·5. Composition:

Selenium 34
Lead 50
Copper 15
Silver.............................. 1—100

Occurs in quantity in the Pearl lode, Minnesota gulch, San Juan county.

RUTILE (p. 58).—Small jet-black tetragonal crystals occur in quartz gangue near the Eureka tunnel, St. Peter's Dome, El Paso county, associated with arfvedsonate and zircon, and closely resembling in form the latter mineral. An analysis by Eakins gave :

Titanic Acid	94·73
Oxyd of Iron	3·77
Silica	1·37
Water	0·71—100 78

SCHIRMERITE (p. 32).—In the Santa Cruz mine, Howard's Fork, San Miguel county, with bismuthinite, chalcopyrite and galena.

EPIDOTE (p. 36).—Good single crystals are found near the Colorado Coal and Iron Co's mine at Calumet.

ANGLESITE (p. 7).—Large masses in the Madonna mine, Chaffee county, often as pseudomorph after galena.

ARSENOPYRITE (p. 40).—Near Ruby camp, Gunnison county. On Mount Wilson, San Miguel county. .

BORNITE (p. 14).—In the Plutus mine, Idaho Springs. In the mines of Red Mountain district, San Juan and Ouray counties. Near Maysville, Chaffee county. Gem mine, Fremont county, with niccolite.

CHALCOCITE (p. 37).—Occurs almost alone, in sandstone with matrix of calcite, in San Miguel county. Massive at Ironton, Ouray county.

HEMATITE (p. 34).—Large, irregular deposit in the Breece mine, Leadville. Large beds near Ashcroft and Carbondale, Pitkin county.

STROMEYERITE (p. 57).—Rarely in the Plutus mine, Idaho Springs. In quantity in the Yankee Girl mine, Ouray Co.

TENNANTITE (p. 8).—Frank Hough mine, Hinsdale Co.

GRAPHITE (p. 27).—In quartz veins 2½ feet thick on branches of Quartz Creek, Gunnison county.

MAGNETITE (p. 34).—Grayback gulch, Costilla county. Titaniferous variety on Iron mountain, Fremont county. Both worked by the Colorado Coal and Iron Company. A pure magnetite in large quantity on Cebolla Creek, and titaniferous magnetite near Snowmass mountain, Gunnison county. In quantity near Hamilton, Park county. Near Ashcroft and Carbondale, Pitkin county.

ACTINOLITE (p. 30).—Light green and bluish green actinolite is found on Mt. Ouray, Chaffee County.

NATIVE SULPHUR (p. 6).—In the Queen of the West mine, Peru District, Summit County, with alabandite and rhodochrosite. In the S. S. mines, Park County. In Routt Co., but the exact locality is not known. In the Mineral Chief lode, Georgetown, with galena.

RHODOCHROSITE (p. 10).—Rich red, transparent crystals in the John Reed mine at Alicante, Lake county. Some cleavage pieces are as pellucid as Iceland spar and show the same strong double refraction. It also occurs opaque, massive and cleavable, enclosing bright cubic crystals of pyrite. Analysis by James B. Macintosh gave:

Protoxyd of Manganese	58.325
Protoxyd of Iron	3.615
Carbonic Acid (by difference)	38.06 —100.00

At the Ule mine, Lake City, opaque flesh-colored crystals, rounded and curved like dolomite, are found on galena. Large masses of white, yellowish and light pink color as stalagmites in the Stevens mine, near Gray's Peak.

HYALITE (p. 41).—Crusts on the hypersthene-andesite of Buffalo Peaks, Chaffee county.

SCHEELITE (p. 27).—In crystals in some of the mines of Baker Park, San Juan county.—(*Endlich*).

CUPRITE (p. 40).—Sweet Home mine, Buckskin, Park Co. North Fork of the Poudre, Larimer county, with native copper and chalcocite.

CELESTITE (p. 25).—On Apishapa Creek.

HESSITE (p. 28).—Mines of Red Cliff, Eagle county. Occasionally met with in the mines of La Plata county.

PETZITE (p. 43).—Hotchkiss mine, Hinsdale county.

SYLVANITE (p. 32).—In quartz veins on Junction Creek, and at the head of the Rio la Plata, La Plata county.

APOPHYLLITE (p. 15).—Endlich mentions its occurrence on Hunt's Peak, Fremont county.

BISMUTHENITE (p. 9).—Near Silverton, in long steel-blue and iridescent needles, associated with hubnerite, tetrahedrite and drusy quartz.

OPAL (p. 40).—Nodules of milk-white opal, some having moss-like markings, on Badger Creek, South Park.

FLUORITE (p. 10).—In veins with quartz in Poughkeepsie gulch, San Juan county. On Kelso mountain, Clear Creek county.

CHRYSOCOLLA (p. 29).—In the Sangre de Christo mountains, Custer and Saguache counties.

BERTRANDITE. *Penfield, Am. Jr. Sc., July,* 1888.
Monoclinic ; crystals generally tabular in habit, transparent, colorless or slightly yellowish. Lustre vitreous. H. 6—7. G. 2·598. Analysis of Colorado mineral gave:

Silica	51.8
Glucina	39.6
Lime	1.0
Water	8.4—100.8

Found on Mt. Antero, Chaffee county, with phenacite, orthoclase, beryl, muscovite and fluorite. Occurs in small thin rectangular blades attached to quartz, and having the shape of a thin slice cut from the side of a cylinder parallel to its axis.

PYRRHOTITE. Magnetic Iron Pyrites.
Hexagonal; commonly massive and amorphous; structure granular. Lustre metallic. Color between bronze-yellow and copper-red, and subject to speedy tarnish. Streak dark grayish-black. Brittle. Magnetic, being attractable in fine powder by a magnet, even when not affecting an ordinary needle. H. 3·5—4·5. G. 4·4—4.68. Composition:

Sulphur	39·5
Iron	60·5—100·0

Occurs in small quantity in the lodes of Needle mountains, San Juan county.

SPHEROSIDERITE. Concretionary Siderite. Carbonate of Iron.
In globular concretions, either solid or concentric scaly, with usually a fibrous structure. Composition :

Carbonic Acid	37·9
Protoxyd of Iron	62·1—100·0

A thin seam of spherosiderite occurs above the coal of the Laramie formation in many parts of the state. In some places it has weathered out in nodules, which are found upon the plains in great quantity. It is found near Trinidad and El Moro in Las Animas county, Walsenburgh in Huerfano county, and Marshall in Boulder county.

HYDROPHANE. Variety of Opal.
Massive, amorphous. Lustre vitreous, frequently subvitreous. Translucent. Color whitish, or light colored. Adheres to the tongue and becomes more translucent or transparent in water. Composition :

Silica	93.00
Alumina	2.00
Water	5.00—100.00

A white and nearly opaque variety of hydrophane, in rounded lumps, with a white chalky or glazed coating, is found in the State, but the locality is a secret. When water is dropped upon it, it first becomes very white and chalky, and then gradually perfectly transparent. This property is developed so strikingly that the name ''Magic Stone'' has been proposed for it, and its use suggested in rings, lockets, charms, etc., to conceal photographs, hair or other objects which the wearer wishes to reveal only when his caprice dictates (*Kunz*).

SAMARSKITE. *Hillebrand, Proc. Col. Sc. Soc., 1888.*
Orthorhombic. Lustre of surface of fracture shining and submetallic.
Color velvet-black. Streak dark reddish-brown. Opaque. Fracture sub-
conchoidal. H. 5.5—6. G. 5.614—5.69.

Occurs in granitic debris on Devil's Head Mountain, Douglas County, in fragments of all sizes up to that of a chestnut,
without crystal form. The following analysis by Hillebrand
shows the composition:

Tantalic Acid	27.03
Columbic Acid	27.77
Tungstic Acid	2.25
Oxide of Tin	0.95
Zirconia	2.29
Protoxyd of Uranium	4.02
Thoria	3.64
Protoxyd of Cerium	0.54
Oxide of Didymium and Lanthanum	1.80
Oxide of Erbium	10.71
Yttria	6.41
Sesquioxyd of Iron	8.77
Protoxyd of Iron	0.32
Protoxyd of Manganese	0.78
Zirconia	0.05
Oxide of Lead	0.72
Lime	0.27
Potassa	0.17
Soda and Lithia	0.25
Water	1.58
	100.31

MEGABASITE. Tungstate of Manganese.
Orthorhombic; usually occurs in fine needles. Lustre vitreous, a little
adamantine. Color brownish-red, clover-brown to yellowish-brown, with
a reddish-brown to hyacinth-red translucency. Streak pale yellowish-
brown to ochre-yellow. H. 3.5—4. G. 6.45.

H. F. Keller gives the following analysis of megabasite
from Bonita Mountain, near Silverton:

Tungstic Acid	74.24
Oxide of Manganese	21.09
Protoxyd of Iron	2.06
Oxide of Copper	0.11
Oxide of Manganese	trace.
Silica	2.13
	99.63

ESSONITE. Lime-Aluminagarnet.
A cinnamon-colored variety of garnet. Composition :

Silica	41
Alumina	23
Oxide of Iron	4
Lime	32—100

Longfellow gulch, Chaffee county, in dark rhombic dodecahedrons, disseminated through chloritic schist. In metamorphosed limestone near Calumet. They have finely polished
faces and beautiful clear color.

WARRENITE. (*New.*) Sulphantimonate of Lead. *L. G. Eakins, Proc·*
Col. Sc. Soc., 1888.
Small acicular crystals forming matted wooly-like masses. Color grayish-
black, occasionally iridescent. Analysis by Eakins:

Lead	39.33
Antimony	36.34
Iron	1.77
Silver, Copper, Manganese	trace.
Sulphur	21.19
Insoluble gangue	.52
	99.15

Found in the Domingo lode, on the ridge between Canon
and Baxter Basin, Gunnison county. Named after E. R.
Warren of Crested Butte.

FREIESLEBENITE. Sulphuret of Silver and Antimony. *Eakins,*
Proc. Col. Sc. Soc., 1888.
Monoclinic; occurs in small, deeply striated prisms. Lustre metallic.
Color and streak light steel gray, inclining to silver white, also blackish
lead gray. Yields easily to the knife, and is rather brittle. Fracture subconchoidal—uneven. H. 2—2.5. G. 6—6.4. Comp.: Sulphur 18.6, antimony 25.9, lead 31.2, silver 24.3.—100. Silver often replaced by lead.

Occurs in a mine on Augusta Mountain, Gunnison Co., in
silicious gangue together with pyrite and sphalerite, and
forms groups of acicular crystals which are elongated prisms,
deeply striated, but too small to be measured.   The analysis
is as follows:

Silver	trace.
Lead	55.52
Iron	trace.
Antimony	25.99
Sulphur (calculated)	18.98
	100.40

ENARGITE (p. 13).—In the mines of Summit district, Rio
Grande county.

JET (p. 45).—On the Trinchera mesa, southeast Colorado.

MOSS AGATE (p. 18).—In nodules on Badger Creek, South
Park.

ORTHOCLASE (p. 12).—Excellent crystals are found in the
slide rock just above the river on Crested Butte Mountain,
Gunnison County.

AMETHYST (p. 23).—Pale colored amethyst is found on Car-
nero Creek, Saguache County. In form they are quite remarkable, the prismatic faces being enormously large, while
the terminal pyramidal faces are very minute. "Phantom
crystals" are often enclosed, and occasionally liquid inclu-
sions with movable bubbles.

Clear Creek County is pre-eminently mineral-bearing, the only tillable land being a few rods in width along the streams. The southeastern portion of the county, however, is valuable only for timber, as mineral veins in that region either do not exist or have not been discovered. The formation is archean, veined with dikes of eruptive rocks, in which porphyry predominates. Heavy-bedded and thinly-laminated gneiss, red and gray granite, schist, syenite, felsitic and hornblendic rocks pass by imperceptible phases from one to the other in inextricable confusion, and the dip of the rocks are as variable as their structural character.

It is somewhat remarkable that so few accessory minerals have been found in the granite and kindred rocks, and equally remarkable that the minerals found in the veins are locally few in variety and seldom crystallized. Taken as a whole, however, Colorado is probably the richest in variety of minerals of any known region. But the number of local minerals might be materially increased were the finders as much interested in learning the nature of the minerals that pass through their hands as in determining their gold and silver value. The following minerals are found in Clear Creek County :

Acanthite (orthorhombic silver glance). — A few small crystals were found in a lot of polybasite crystals from the Emma mine.

Albite (soda feldspar).—A constituent of some of the granitic rocks, particularly on Fall River.

Amethyst.—A few specimens of good color have been found in a lode near Dumont and on Chicago Creek.

Anglesite (sulphate of lead).—Aggregations of small crystals in the Freeland mine. Small crystalline masses in the Troy mine, Georgetown, intimately associated with cerussite.

Aragonite (carbonate of lime).—Large beautiful specimens of coralloidal form were found in the Marshall tunnel a number of years ago. A shaft near the Argentine lode, on Leavenworth Mountain, furnishes yellow tinted specimens.

Argentite (silver glance).—In small quantities in nearly all the silver mines. Masses weighing two to three pounds have been found in the Colorado Central and Saxon mines. Crystals rare and quite small.

Arsenopyrite (arsenical pyrites).—Occasionally found in the veins about Empire, Idaho and in Geneva District.

Autunite (phosphate of uranium and lime).—Small greenish-yellow tabular crystals in the Peabody lode, near Georgetown, and in a lode on Chicago Creek.

Azurite (blue carbonate of copper.)—Occurs as an incrustation near the surface of many of the lodes.

Barite (heavy spar).—Of frequent occurrence in some of the veins. Clusters of amber-colored crystals in the 7:30 mine. In the Little Giant mine in thin white crystals. Small tabular crystals in the Oneida and many other lodes.

Biotite (black mica).—A constituent of much of the granite and eruptive rocks.

Bornite (purple copper ore).—Of frequent occurrence in the gold mines. Handsome specimens in the Neath.

Calcite (carbonate of lime).—Occurs in small quantities in many of the mines. Nail-head crystals in the Terrible.

Cerargyrite (horn silver, chloride of silver.)—Thin crusts on the surface ores from some of the lodes on Democrat and Lincoln Mountains.

Chalcanthite (blue vitriol).—Forms on the walls of the Whale Tunnel, near Idaho.

Chalcedony.—Beautiful botryoidal specimens have been found in the mines at Idaho. Occasionally met with in the Little Giant and other veins.

Chalcopyrite (copper pyrites).—Of general occurrence but in less quantity than pyrite, and carries more gold and silver. A few fine crystals, usually studded with minute crystals of tennantite, have been found in the Freeland mine.

Chrysocolla (silicate of copper).—In the Champion lode, near Idaho.

Dolomite (magnesian limestone).—Of frequent occurrence in the veins in Ohio District. Occasionally found in other lodes.

Enargite (arsenical sulphide of copper).—A small amount is found in the surface ores of the Centennial mine, associated with native copper.

Epidote.—Massive and crystallized in the granitic rocks between Georgetown and Green Lake. Very good crystal on Floyd Hill.

Fibrolite.—Inferior fibrolite of a grayish-green color occurs in the granite on Sherman Mountain.

Fire Opal.—A very handsome opal showing bluish and yellowish flames, and greatly resembling some Mexican opals, is found in a felsite rock in Gilson Gulch, near Idaho.

Fluorite (fluor spar).—Impure massive fluorite of deep purple color is found in some of the lodes on McClellan Mt.

Galenite.—The most abundant mineral in the silver mines. Good crystals in the Dunderberg, Terrible and Little Giant mines. In many of the mines it occurs in all grades from coarse cubes to extremely fine granular, and sometimes lamellar. The percentage of silver it carries varies from a few ounces to several hundred ounces per ton.

Garnet.—Large brown garnets are occasionally found imbeded in granite on Griffith Mountain.

Gibralter Rock.—A ledge of this variety of limestone is located near Idaho Springs. When polished it is quite handsome.

Hematite.—Ocherous hematite is found near the surface of many of the veins.

Hornblende.—Of general occurrence in the granitic rocks. An essential constituent of syenite, and is frequently found massive. No good crystals have been found.

Hornstone.—Of frequent occurrence in the mines, forming part of the gangue rock.

Kaolinite (porcelain clay).—Masses showing crystalline structure in the Silver Mountain lode, near Empire. Occurs in many other lodes in small quantity.

Linarite (cupreous anglesite).—A mineral having the general characteristics of linarite is found at the surface of some of the lodes on Red Elephant Mountain.

Magnetite (magnetic iron).—Crystals of magnetite are of frequent occurrence in granite. The granite quarried at Brownville is liberally sprinkled with crystals of this mineral.

Malachite (green carbonate of copper).—Occurs as a stain or incrustation at the surface of many of the veins.

Mesitite.—Found in small quantities at the surface of the Colorado Central vein.

Micaceous Iron (hematite).—Quite abundant between Clear Creek and Mill Creek.

Molybdenite.—Small masses in the Louisiana lode, Sherman Mt., in the Colorado Central mine, in the Moline Tunnel, in several localities on Red Elephant Mt., and in Daily District.

Muscovite (common mica).—Common in archean rocks. Sometimes found in large masses. Very good crystals are found in a mica claim near Graymont.

Native Copper.—Small dendritic scales in the Centennial and McClellan mines; wires and small masses in the Pittsburg mine, Empire. Loose masses have been found in the soil about Idaho.

Native Gold.—In placers and in veins along the creek from Idaho to Empire, on Fall River and in Argentine District. Many of the silver mines also carry an appreciable amount of gold.

Native Silver.—In leaf and wire form in many of the mines throughout the county. Nuggets in the talc bed on Fall River.

Niccolite (arsenical nickel).—Occurs sparingly in the Rosa lode, Dumont.

Orthoclase (common feldspar).—One of the most common rocks throughout the county and occurs in immense masses as well as a constituent of nearly all country rock.

Polybasite.—Next to gray copper polybasite is the most abundant rich silver ore found in the county. Crystals are quite abundant in the Emma mine, and some have been found more than an inch in their longest diameter. A few crystals are met with in the Colorado Central. It usually occurs massive, when it is difficult to distinguish from gray copper.

Proustite (light ruby silver ore).—Small masses and minute translucent crystals in the Colorado Central mine.

Pyrargyrite (dark ruby silver ore).—Of general occurrence in the silver mines but usually in small quantity. Probably more is found in the Colorado Central than in any other mine. Crystals have never been noticed.

Pyrite.—Almost universally in the mines. Particularly abundant at Empire and Idaho, and in the "gold belt" lodes at Georgetown. Usually massive; crystals quite small and rarely vary from the cube. Carries gold and a little silver.

Pyromorphite (phosphate of lead).—Sparingly in some of the mines on Democrat Mountain.

Quartz.—In large masses at the eastern end of the county. Forms part of the gangue rock of the mines. Crystals frequent but usually small.

Rhodochrosite (carbonate of manganese).—Some very pretty crystals are found in the Danube lode, near Idaho Springs.

Rose Quartz.—Beautiful rose-tinted masses on Floyd Hill. Crystals in the Little Giant mine.

Schirmerite (bismuth silver ore).—In small masses in the mines of Geneva District.

Siderite (carbonate of iron).—Of frequent occurrence, both massive and crystallized, in the mines in the Trail Creek region. Occasionally in other mines.

Sphalerite (zinc blende).—Abundant, especially in the Georgetown and Silver Plume district. Varies in color from amber to black. Large clusters of black crystals in the Maine mine, some crystals three inches in diameter. Usually carries but little silver.

Stromeyerite.—A variety carrying a little more than one per cent of silver is found in small quantities in the Algonquin, Silent Friend and other lodes on Saxon Mountain. In large quantities in the Plutus mine, mixed with bornite, and carrying 25 per cent of silver. In small acicular crystals in the Little Giant mine, and carrying about 50 per cent of silver.

Talc.—An extensive bed near the head of Fall River which is being sluiced by the Alice company for the gold it contains.

Tennantite (arsenical sulphide of copper).—Clusters of highly polished crystals from microscopic size to ⅓ of an inch in diameter in the Freeland mine.

Tourmaline.—Inferior tourmaline is found on Baker Mountain and near Green Lake.

Tetrahedrite (gray copper).—The most abundant rich silver ore. Gray copper from the mines in the vicinity of Georgetown and Silver Plume carries as high as 30 per cent of silver, while that from the mines east of Georgetown and in the vicinity of Idaho Springs, and in Geneva District rarely contains more than one per cent of silver. A few coarse crystals have been found in the Geneva District mines.

Uraninite (pitchblende).—Small masses have been found in the Jo Reynolds mines.

GILPIN COUNTY MINERALS.

Albite (soda feldspar).—A constituent of some of the archean rocks.

Amethyst.—Small beautiful crystals of good color have been found in the gravel in Nevada.

Arsenopyrite (arsenical pyrites).—Occasionally found in small masses in the Kansas, Illinois, Burroughs, California and other mines.

Azurite (blue carbonate of copper.)—Occurs as an incrustation or stain at the surface of many of the mines.

Barite (heavy spar).—Of frequent occurrence in some of the mines, and occasionally crystallized.

' *Biotite* (black mica).—A constituent of much of the granite and eruptive rocks.

Bornite (purple copper ore).—Of general occurrence in the mines in greater or less quantities. Crystals rare..

Calcite (carbonate of lime).—Occurs in small quantities in many of the mines.

Chalcanthite (blue vitriol).—Near Black Hawk, in a deposit. In many of the old mine dumps about Central.

Chalcedony.—Small masses occasionally met with in many of the mines.

Chalcopyrite (copper pyrites).—Of general occurrence and one of the principal ores of gold. Crystals frequent.

Covellite (indigo copper).—In grains and as a powder in Pewabic, Gunnell, Sapp and other lodes near Central.

Enargite (arsenical sulphide of copper).—Massive and finely crystallized in the Powers mine, Russell District.

Epidote.—Massive and crystallized in the granitic rocks throughout the county.

Float Stone.—Snow-white masses showing gold have been taken from the Mammoth mine, Central.

Fluorite (fluor spar).—Met with occasionally in some of the veins.

Galenite.—Abundant in many of the mines. Fine crystallized specimens have been found in the Glennan and Running mines, Black Hawk, the Delaware and Calhoun lodes, Russell District, the Mount Desert and Forks lodes at Nevada. Crystal two inches across have been found in the Glennan.

Garnet.—Iron garnets are quite abundant in the vicinity of Central.

Hornblende.—Of general occurrence in the archean rocks.

Hornstone.—Of frequent occurrence in the mines, forming part of the gangue rock.

Kaolinite (porcelain clay).—In greater or less abundance in all the mines.

Magnetite (magnetic iron).—Crystals of magnetite are of frequent occurrence in granite. Octahedral crystals on Gunnell Hill.

Malachite (green carbonate of copper).—Occurs as a stain or incrustation at the surface of many of the veins.

Melanterite (copperas, green vitriol).—In the dumps of many of the old mines.

Molybdenite.—Small masses in the Leavitt lode.

Muscovite (common mica).—Common in archean rocks. Sometimes found in large masses.

Oligoclase (aventurine feldspar).—A constituent of much of the archean rocks.

Native Copper.—Small dendritic coatings in a number of the lodes. Masses weighing several pounds have been found in the Narraganset and Gregory mines.

Native Gold.—In placers and in the veins. Beautiful microsopic crystals in the Gunnell, Gregory, Bates and other mines. Strings of minute crystals, forming chain-like wires, are frequently found.

Native Silver.—In leaf and wire form in the Coaley and other mines.

Orthoclase (common feldspar).—One of the most common rocks throughout the county and occurs in immense masses as well as a constituent of nearly all of the rocks. Crystals in porphyry on Gregory Hill.

Pyrite (iron pyrites).—Abundant, massive and crystallized, in all the lodes, and composes the greater part of the gold ore. Beautiful crystallized specimens and radiated and botryoidal masses common in many of the mines. Cubes from one to three inches across have been taken from the Kingston, Hill House and S. P. Chase lodes in Russell District. A lode near the head of Virginia Canon has furnished a great many modified crystals.

Rose Quartz.—In small masses in the vicinity of Central and other localities.

Quartz.—Common in large masses.   Forms part of the gangue rock in the mines.   Crystals frequent but usually small.   An essential constituent of granitic rocks.

Siderite (carbonate of iron).—Of frequent occurrence, both massive and crystallized, in the mines.

Rock Crystal.—Doubly terminated limpid crystals common in the gravel on Quartz Hill.   The crystals are small but quite perfect.

Sphalerite (zinc blende).—Abundant in many of the lodes. Clusters of tetrahedral crystals in the Coaley lode, Black Hawk.   Crystallized in the Delaware and Calhoun lodes.

Talc.—In small quantities in nearly all the mines.

Tourmaline.—Black crystals near Nevada.

Tetrahedrite (gray copper).—Not unfrequently found in a number of mines, sometimes crystallized.

Uraninite (pitchblende).—Abundant at one time in the Wood lode, Leavenworth Gulch.   A few imperfect crystals.

Uraconite.—Occurs as a yellow coating on pitchblende in the Wood lode.

Cerussite (carbonate of lead).—Of rare occurrence at the surface of some of the veins.

Lepidomelane (iron potash mica).—A constituent of some of the granite.

NEEDS VERIFICATION.

Aurichalcite.—Jones lode, near Nevada.

Allophane.—Cincinnati lode.

Calamine.

Glockerite.—On old dumps.

Goslarite.—Wood lode, Leavenworth Gulch.

Greenockite.—Jones and Dallas lodes.

Jarosite.—Wood lode.

Lievrite.—Below Black Hawk.

Magnesite.—Bobtail, Running and Pewabic lodes.

Marcasite.

Native Sulphur.—Jones lode, Nevada.

Plumbogummite.—Dallas lode, Black Hawk.

Pyrrhotite.

Smithsonite.—Jones lode, Nevada.

Uranochalcite.—Wood lode.

Willemite.—Jones lode, Nevada.

Wulfenite.—Gunnell Hill.

Zincite.—Jones lode, Nevada.

Zippeite.—Wood lode.

GRAND COUNTY MINERALS.

Albite (soda feldspar).—A constituent of some of the archæan rocks.

Agate.—Abundant along Williams River; on the ridge between Hot Sulphur Springs and Corral Creek ; on the mountains at the mouth of Willow Creek.

Agatized Wood.—Abundant on the ridges north of Willow Creek. Also between Hot Springs and Corral Creek.

Argentite (silver glance).—In small quantities in the mineral veins in Campbell district.

Azurite (blue carbonate of copper.)—Occurs as an incrustation in copper veins in the Park Range at the head of a fork of the North Plate River.

Barite (heavy spar).—Reported to occur in extensive beds on Vasquez Creek.

Basanite (touch stone).—Sparingly along Willow Creek.

Biotite (black mica).—A constituent of much of the granite and dike rocks of the mountains and eruptive rocks of Middle Park.

Bismutite (carbonate of bismuth).—Found near Cummuns City, North Park. Needs verification.

Brown Coal (lignite).—Extensive veins in North Park and exists at various points on the divide between North and Middle Parks.

Byerite.—A name given to the coal found in Middle Park. It belongs to the class caking bituminous, and resembles albertite in the large amount of gas and tarry oil yielded. It melts and intumesces when heated.

Carnelian.—Sparingly on the ridge south of Willow Creek.

Cerussite (carbonate of lead).—Frequently found in small masses near the surface of mineral veins.

Chalcedony.—Abundant on the ridges on both sides of Willow Creek. Common on the ridge between Hot Springs and Corral Creek.

Chalcocite (copper glance, sulphuret of copper).—In copper veins in the Park Range.

Chalcopyrite (copper pyrites).—Of general occurrence in small quantities in the mines.

Chrysoprase.—Sparingly in geodes on upper Grand River.

Cuprite (red oxide of copper).—In copper veins in the Park Range.

Cyanite.—Occurs in the quartzite of Medicine Bow Peak.

Epidote.—Massive and crystallized in the granitic rocks.

Galenite.—Abundant in many of the mines. Massive veins near the head of Troublesome Creek.

Heliotrope (bloodstone).—Fine specimens occasionally met with in the vein of green jasper on the hill at the junction of Willow Creek and Grand River.

Hornblende.—Of general occurrence in the archæan rocks.

Hypersthene.—A constituent of the feldspar basalt from Grand River, above Hot Springs.

Jasper.—Veins of green, red and yellow jasper on the hill at the junction of Willow Creek and Grand River. Ribbon jasper is quite common at this locality. Red and yellow jasper is more or less abundant on the ridges along Willow Creek, in the vicinity of Hot Springs, and in the streams.

Lepidomelane (iron potash mica).—A constituent of some of the granite.

Magnetite (magnetic iron).—An accessory mineral of much of the granite.

Malachite (green carbonate of copper).—Occurs as a stain or incrustation at the surface of many of the copper veins in the Park Range.

Moss Agate.—Quite abundant on the south side of Williams River, three or four miles from the Grand.

Mottled Agate.—On Williams River; in the vicinity of Hot Springs; Willow Creek, and numerous other localities.

Muscovite (common mica).—Common in archæan rocks, frequently occurring in large masses.

Native Bismuth.—Occurs in small particles in bismutite at Cummuns City. The find is not satisfactorily verified.

Native Gold.—In the alluvium of Willow Creek.

Native Silver.—In leaf and wire form in some of the silver-bearing veins.

Oligoclase (aventurine feldspar).—A constituent of much of the archæan rocks.

Onyx.—Sparingly with chalcedony on the ridge south of of Willow Creek.

Orthoclase (common feldspar).—The most abundant constituent of the archæan rocks.

Plasma.—Occurs in geodes found on upper Grand River.

Pyrargyrite (dark ruby silver ore).—In small quantities in the lodes of Campbell district.

Pyrite (iron pyrites).—More or less abundant in the mineral veins.

Pyroxene.—Next to the feldspars, pyroxene is the most universal constituent of the igneous rocks.

Quartz.—Common, in crystal, small masses and a constituent of metamorphic and sedimentary rocks.

Sardonyx.—Fine specimens occasionally met with on the ridge south of Willow Creek, a mile or two from the Grand.

Stibnite (antimony glance).—Float masses of stibnite and quartz, two feet in thickness, are found between the head of Troublesome Creek and Lost Lake.

Sphalerite (zinc blende).—Abundant in many of the lodes.

Tetrahedrite (gray copper).—Not unfrequently found in the mines at Lulu. .

JEFFERSON COUNTY MINERALS.

Actinolite.—Columnar masses of green color near Bergen ranch on Bear Creek, and at the entrance of Coal Creek canon.

Agate.—Mottled agate common on Green Mountain.

Agatized Wood.—Abundant on Table Mountains.

Alabaster.—Massive white and mottled alabaster in Jurassic shales near Morrison.

Albite (soda feldspar).—A constituent of some of the archæan rocks.

Allophane.—Forms a thin bluish crust on limonite in veins near Bergen ranch, Bergen Park.

Amazonstone.—On Elk Creek and near the mouth of Tarryall Creek, with smoky quartz and gothite.

Analcite.—Pure white or transparent crystals in the basalt of North Table Mountain, associated with other zeolite minerals. Specially abundant on the east side of the mountain. Small crystals abundant on South Table Mountain.

Apophyllite.—Well developed crystals of prismatic habit on North Table Mountain. The larger crystals are usually terminated by a number of small pyramids.

Aquamarine.—Small inferior specimens on Tiffany ranch, Bear Creek.

Aragonite.—Occurs in a vein at Golden Gate.

Asphaltum (bitumen, mineral pitch).—Forms crusts on sandstone near Morrison.

Augite.—An essential constituent of the augite-andesite rocks of Green and Table Mountains.

Azurite (blue carbonate of copper.)—Occurs as an incrustation and minute crystals in the copper veins, Bear Creek.

Beryl.—Abundant on Bear Creek and in Bergen Park. Crystals vary in size from small ones to those two feet in diameter. Translucent and good color specimens rare.

Biotite (black mica).—A constituent of much of the granite and dike rocks.

Bismutite (carbonate of bismuth).—In the Bismuth Queen vein on Guy Hill.

Bismuthinite (sulphuret of bismuth).—Associated with bismutite in the Bismuth Queen lode, Guy Hill.

Bole (iron clay).—Occurs as a dark brown clay on South Table Mountain.

Brown Coal (lignite).—Extensive veins at Golden, on Ralston Creek, and other localities.

Calcite (carbonate of lime).—Large masses of cleaveable white calcite on Bear Creek.

Cairngorm (smoky quartz.)—On Elk Creek and near the mouth of Tarryall Creek, with amazonstone and gothite.

Carnelian.—Small pieces of red and yellow carnelian on Green Mountain.

Chabazite.—Abundant in small crystals in the basalt of North Table Mountain.

Chalcedony.—Sparingly on Green Mountain and vicinity.

Chalcedonyx.—Sparingly between the mountains and Denver.

Chalcocite (copper glance, sulphuret of copper).—In copper lodes on Bear Creek, in Bergen Park, and near Golden.

Chalcopyrite (copper pyrites).—In the mineral veins on Bear Creek and in Bergen Park.

Chlorophane (green fluor spar).—Abundant in many veins on Bear Creek.

Chrysocolla (silicate of copper).—Sparingly in the copper veins on Bear Creek and in Bergen Park.

Columbite.—In small masses on Turkey Creek. Analysis shows the mineral to contain a small amount of tin and 11 per cent of manganese.

Cuprite (red oxide of copper).—In copper veins in Bergen Park and on Bear Creek.

Enstatite.—An accessory mineral of a dike of diabase at Morrison.

Epidote.—Massive and crystallized in the granitic rocks. Quite pretty crystals are found on Bear Creek.

Fluorite (fluor spar).—Massive, purple, green and white, in veins on Bear Creek. Purple near the mouth of Tarryall. Massive on Cub Creek.

Galenite.—Abundant in many of the veins on Bear Creek.

Garnet.—Iron garnets are quite common in Bergen Park.

Gothite.—On Elk Creek and near the mouth of Tarryall Creek, with amazonstone and smoky quartz.

Gypsum.—Abundant in Jurassic shale near Morrison.

Hematite.—An earthy red hematite at Morrison.

Heulandite.—Small colorless tabular crystals in the augite-andesite of Green Mountain.

Hornblende.—Of general occurrence in the archæan rocks.

Iceland Spar.—Small yellowish masses in the amygdaloid of the Table Mountains.

Jasper.—Red and yellow jasper in the Green Mountain region.

Kaolinite (porcelain clay).—Extensive beds at Golden.

Laumontite.—Occurs as a reddish-yellow sand-like material and in compact masses in the basalt of North Table Mt.

Lepidomelane (iron potash mica).—A constituent of some of the granite.

Levynite.—Of rare occurrence in small white and colorless crystals associated with other zeolites on North Table Mt.

Limonite.—Small masses in the veins of Bergen Park.

Magnetite (magnetic iron).—An accessory mineral of much of the granite. Loose crystals in the soil on Bear Creek and in Bergen Park.

Malachite (green carbonate of copper).—Occurs as a stain or incrustation at the surface of many of the copper veins in Bergen Park and on Bear Creek.

Mesolite.—Occurs as exceedingly delicate needles in the basalt of North Table Mountain.

Microcline.—Near the mouth of Tarryall Creek, with amazonstone, smoky quartz, gothite and fluorite.

Mineral Resin.—Small specimens of an undetermined variety of resin in the soil on South Table Mountain and in the coal veins.

Molybdenite.—Occurs as an associate of bismuth minerals on Guy Hill.

Muscovite, (common mica).—Common in archæan rocks, Large deposits on Deer Creek, where some mica mining has been done.

Native Alum.—In Jurassic shale on Bear Creek near Morrison, and between Morrison and Platte Canon.

Native Copper.—In considerable quantities in copper lodes near Golden. Small specimens in the lodes on Bear Creek and in Bergen Park.

Native Gold.—In the alluvium of Clear Creek.

Native Sulphur.—In small crystals. Exact locality not known.

Natron (carbonate of soda).—As crusts and in solution in the waters of the lakes between Turkey and Bear Creeks.

Natrolite.—Occurs as delicate prisms sparingly deposited on analcite or associated with that mineral in the basalt of South Table Mountain. Less frequently on North Table Mt.

Oligoclase (aventurine feldspar).—A constituent of much of the archæan rocks.

Orthoclase (common feldspar).—The most abundant constituent of the archæan rocks.

Petroleum.—Oozes from Cretaceous sandstone at Morrison.

Plagioclase.—An essential constituent of diabase from Morrison, in which it occurs in narrow prisms.

Ptilolite (new).—A new zeolite described by Whitman Cross. Occurs in small white tufts in the augite-andesite of Green Mountain. Sparingly with other zeolites on the Table Mts.

Pyrite (iron pyrites).—Small quantities in the veins of Bergen Park and on Bear Creek.

Pyroxene.—An essential constituent of diabase from a dike near Morrison.

Quartz.—Large veins and masses of white quartz in Bergen Park, on Bear Creek, and in the Guy Hill region.

Rock Crystal.—Colorless quartz in crystals abundant on the divide between Deer and Elk Creeks.

Rose Quartz.—In large masses in Bergen Park.

Satin Spar (var. gypsum).—Abundant in Jurassic shales near Morrison.

Scolecite.—Occurs sparingly in spheres with a radiate structure in the basalt of North Table Mountain.

Selenite (gypsum).—In Jurassic shale near Morrison.

Silicified Wood.—On South Table Mountain. Silicified palm wood abundant between Golden and Denver.

Stilbite.—Sparingly in small clear crystals with other zeolites on North Table Mountain.

Thenardite (sulphate of soda).—Occurs in quantity at Burdsall's lake near Morrison.

Thomsonite.—In the basalt of North Table Mountain in minute rectangular blades which are placed like a closed fan, and in spherical concretions having a radiate structure.

Tourmaline.—Fine black crystals in quartz on Bear Creek. Near the head of Ralston Creek.

Jefferson County is also accredited with the following minerals, but their occurrence needs verification :

Aluminite, Mt. Vernon.

Antrimolite, Table Mountains.

Egeran, Gennessee ranch on Bear Creek.

Idocrase, Bear Creek.

Pyrrhotite.

Wavellite, Table Mountains.

Wheelerite, in the coal.

Topazolite, Malachite lode, Bear Creek.

www.ingramcontent.com/pod-product-compliance
Lightning Source LLC
Chambersburg PA
CBHW021423090426
42742CB00009B/1226